Tiny
Quantum
Giant
Revolution

Tiny
Quantum
Giant
Revolution

Ching-Ray Chang
Meng-Chien Wang

National Taiwan University, Taiwan

World Scientific

NEW JERSEY · LONDON · SINGAPORE · BEIJING · SHANGHAI · HONG KONG · TAIPEI · CHENNAI · TOKYO

Published by

World Scientific Publishing Co. Pte. Ltd.

5 Toh Tuck Link, Singapore 596224

USA office: 27 Warren Street, Suite 401-402, Hackensack, NJ 07601

UK office: 57 Shelton Street, Covent Garden, London WC2H 9HE

Library of Congress Control Number: 2024938784

British Library Cataloguing-in-Publication Data
A catalogue record for this book is available from the British Library.

ISBN 978-981-12-8700-8 (hardcover)
ISBN 978-981-12-8740-4 (paperback)
ISBN 978-981-12-8701-5 (ebook for institutions)
ISBN 978-981-12-8702-2 (ebook for individuals)

For any available supplementary material, please visit
https://www.worldscientific.com/worldscibooks/10.1142/13697#t=suppl

Typeset by Stallion Press
Email: enquiries@stallionpress.com

Preface

Understand the universe through quantum and change the world with technology.

As humans have been active for a long time on this planet that was accidentally formed near the sun, our essence of historical knowledge has been extracted from the life experiences of countless people. The philosophical thinking only with brain and mathematics with paper and pencil were naturally first appeared, and then driven by the desire through alchemy and seek immortality, humans synthesized chemistry using the materials around them. The apple falls to the ground naturally after it ripens, and this wonderful accident unexpectedly unlocked deep wisdom in Newton's mind and solved the physical mystery of the laws of the universe in one fell swoop. In the competition for knowledge among countless wise men, the mysterious veil of the universe has been gradually unveiled. The rapid progress of science and technology in the 20th century has had a huge impact on the world; it has caused an astonishing growth in the population, with the global population reaching 8 billion in 2023, and extended life expectancy to the 80s. This change has far exceeded the normal speed of Darwinian natural evolution.

Galileo opened Pandora's Box and released modern science on Earth. Newton's three laws regulated and elaborated the behaviour of the classical world, and masters such as Heisenberg and Schrödinger opened up the scientific paradigm of quantum. The influence of quantum theory on modern life has been indisputable since the 20th century, with quantum theory now comprehensively integrated into everyone's lives. You may not know the concepts of energy levels, but the mobile phones you use every day rely on semiconductors to work. You may not understand the tunnelling effect, but flash memory is an indispensable tool. Science seems so difficult and far away to most people, but when we look back, we find that science is actually everywhere. Human beings have long been accustomed to living under the huge influence of modern technology. This book will provide you with a basic understanding of quantum computers, quantum communication, and quantum sensing, and the concepts of quantum superposition, entanglement, and measurement in the great Hilbert mathematical space. Therefore, this book aims to build up the necessary literacy and knowledge for quantum citizens in the future century.

Quantum science was founded in Europe in the early 20th century and then gradually matured and promoted the understanding of the universe. Many concepts of quantum science are indeed significantly different from the classical macroscopic world. For example, the concepts of quantization and uncertain probability have not only caused great changes in science and technology but have also caused much discussion in humanities and philosophy. The first Industrial Revolution was accompanied by the emergence of the field of engineering, the second brought about the field of electrical engineering, and the third digital revolution led to the field of information. It is now obvious that this quantum revolution has led to the emergence of the field of quantum science and technology. In 2018, the European Union and the world officially launched the "Second Quantum Revolution." The new quantum technology developed by major companies will rapidly promote human civilization in the next few decades, and will once again have a strong impact on new ideas in the fields of humanities and philosophy. Quantum technology can truly develop comprehensively through the integration of science, technology, engineering, art, and mathematics (STEAM). In addition to the need for breakthroughs in problems within the field of physics, application changes in cross-field disciplines are also an important achievement of the quantum technology revolution. Physics has a complete basic theory and also provides an effective analysis method. The combination of physics and chemistry leads to a new era of quantum and materials chemistry. The various emerging fields of research in life sciences that are using physics, such as molecular biology, genetic cryptography, and protein folding, are growing in scope and becoming more graphic through physical tools. The combination of physics, social and economic theory, and nonlinear mechanics transforms complex financial figures into systematic and organized physics. The 21st-century quantum technology symphony of STEAM will eclipse the original knowledge solos of the 20th century.

This book outlines the importance of the "Second Quantum Revolution," introduces quantum computers, quantum communications, and quantum sensors, and then provides a framework for the emergence of the quantum Internet of Things. What basic quantum literacy should modern citizens have in this era? The "Second Quantum Revolution," where quantum knowledge and engineering technology are once again combined, will provide faster, more effective, and more sensitive quantum facilities to accelerate cross-field exploration, and will also make human life more comfortable and convenient than ever before. In the "First Quantum Revolution" in the 20th century, humans learned quantum science from nature and used existing materials to make quantum components. In the "Second Quantum Revolution" in the 21st century, humans further used quantum science to construct quantum engineering. We now make materials and components that are not found in nature and assemble new quantum machines to benefit mankind! This is the stage of a glorious quantum era, which is a hundred times more brilliant than the past classical physics era.

History needs to be faced honestly, culture is accumulated over time, and science is also a product of human activities, so it cannot be achieved quickly. A brave review of history can illuminate the direction for the next step. Quantum knowledge appeared in Europe in the early 20th century, and has been carried forward in the United States since then. In this decade of the 2020s, the post-Silicon Valley era has begun, and the clarion call for a decisive battle in the Quantum Valley has been sounded. Will the newly appearing quantum centres remain in the US or move to Asia? Now that every country in the world is investing heavily in scientific and technological research of participating in the world's quantum arena. The world's "Second Quantum Revolution" is like the beginning of a dinner party, and the costumed protagonists are about to appear one after another. Quantum technology is saying to everyone, "Do not escape," because giving up quantum means giving up the future. In this book, I tried to transform difficult quantum concepts into common sense. Accept quantum, use quantum, and you will become a modern quantum citizen. This is a grand event that will never happen again in billions of years. There is no time to hesitate. What will the quantum era be like? Is there a wider space and time outside the universe? Prepare yourself first to become a modern citizen of the "Second Quantum Revolution." Only then will you have the opportunity to participate in and control the future, and be forever entangled with quantum citizens in the modern world!

I would like to thank Professor Long Guilu of Tsinghua University for his enthusiastic help in proofreading some chapters of this English version of the book. In the end, I do not know whether I should be grateful to circumstances or not as it was the isolation period in Taipei during the COVID-19 epidemic that gave me the time to write this book. Otherwise, it is difficult to imagine will this book complete in time.

Ching-Ray Chang
November 2023

Contents

Contents

https://doi.org/10.1142/9789811287015_0001

Chapter 1

Quantum Supremacy Has Arrived

Quantum era is approaching, believe in quantum, learn quantum, and use quantum now.

— Ching-Ray Chang (1957–)

The history of the universe is, in effect, a huge and ongoing quantum computation.

— Seth Lloyd (1960–)

1.1 The technological revolutions driven by knowledge applications

1.1.1 *Introduction*

Nearly a century ago, 29 of the world's top scientists gathered in Brussels for a quantum concept that would have stumped even Albert Einstein. Now, this bizarre concept has been transformed into disruptive technologies, which are changing our lives and making quantum knowledge the necessary common sense for modern citizens.

In the era of rapidly changing science and technology, continuous innovation is an important driving force for progress. Quantum theory and quantum mechanics that were developed in the late 19th and early 20th centuries are significant breakthroughs in human exploration and mastery of the microscopic world. After the Industrial Revolution involving the steam engine in the 18th century, science and technology began to have a major impact on human life and thinking. From the 20th century to the present, the applications of acoustic, optics, mechanics, and electronics have further promoted human life to the most glorious and comfortable situation in history. The second quantum revolution and industrial innovation initiated by quantum technology will make life more convenient in the future.

In recent years, China has made rapid progress in science and technology, surpassing other countries in the world in some specific fields. Not only did China's moon mission impress the world but *From the Great Wall to the Great Collider* by Steve Nadis and Shing-tung Yau also powerfully conveyed the fact that China's technological and economic power has indeed developed in response to the call of the world in a way that lives up to history's glorious standards. Presently, countries worldwide are actively investing in the research and development of quantum technology and talent cultivation. China and the United States are competing for leadership in the quantum field, and emerging countries are also trying to progress in this field. In the ongoing development of quantum technology, whoever can do the following will establish technological advantage and take the leading position in the field of quantum technology in this century: grasp the primary value of quantum technology; analyze quantum development trends in depth; combine fundamental research, local technology, and quantum engineering; put theoretical results into practical use; strengthen the integration and innovation of industry and academia; and cultivate quantum interdisciplinary talents.

What is so amazing about quantum mechanics? We can divide the universe into two main classes: the macroscopic world, which obeys classical physics, and the microscopic world, which obeys quantum mechanics. They have very different properties and operating laws. Newtonian mechanics is one of the great discoveries of classical physics. It accurately describes the macroscopic world. Newtonian mechanics can calculate all states and physical quantities of objects, whether the rotation of a star, the launch of a rocket, or the throw of a baseball by a major-league pitcher, with precise knowledge of the past, present, and future. At the beginning of the 20th century, scientists found that the structure of an atom can be divided into the nucleus and the electrons outside the nucleus, with the electrons revolving around the nucleus. But, do the electrons move in the same way as the earth moves around the sun? Noticing that atoms in the microscopic world behave very differently from planets in the macroscopic world, physicists began to develop new quantum theories to describe microscopic physical phenomena. With the collective efforts of many scientists, quantum mechanics was gradually established. It was found that not only is the electron not a particle but its orbiting domain is not the elliptic orbit of a planet. Specifically, electrons appear as probability waves in the orbital domain outside the nucleus, and the position and the momentum of a particle cannot be determined precisely at the same time, which is known as the famous Heisenberg uncertainty principle. In the quantum world, objects always appear as probability waves, which is one of the major differences between the classical and quantum worlds. Microscopic particles, such as atoms, molecules, and photons, all move in accordance with quantum mechanics. Quantum mechanics is the basis of modern technologies and equipment, such as transistors, lasers, and high-temperature superconductors.

Quantum technology is a perfect combination of physics and engineering. The "First Quantum Revolution" brought to the world technologies that changed our modern life, such as semiconductors, optoelectronic devices, smartphones, and the Internet of Things. During the "Second Quantum Revolution," the properties of quantum mechanics, such as quantum entanglement, quantum superposition, non-locality, no-cloning, and quantum measurement, were applied in quantum computers, quantum algorithms, quantum sensors, quantum communication, quantum computation, and quantum simulation. The "Second Quantum Revolution" broke through the sensitivity and functional limits of the traditional technology of the "First Quantum Revolution," thus opening up a new frontier of quantum technology in the future. Before introducing the "Second Quantum Revolution," let us review the history of modern technological development.

1.1.2 *Waves of the industrial revolution*

The "First Industrial Revolution" (1760–1830) started in 1760 when James Watt improved the steam engine. It was a technological revolution in which machines replaced human and animal power, and mass production in factories replaced individual production. The GDP per capita rapidly increased around the Industrial Revolution, showing that technology had indeed boosted productivity.

The "Second Industrial Revolution" (1850–1920) moved us from the steam age to the electric age. During this period, Western Europe, the United States, and Japan made significant innovations in various fields and developed modern electric, chemical, and petroleum industries. Michael Faraday discovered electromagnetic induction, which is the foundation of electric motor technology. Because of the invention of electric generators and motors, combined with the mature long-distance transmission technology of alternating current, electricity was rapidly popularized and widely used in production and life. The advent of the internal combustion engine led to the growth of the automobile and aircraft industries as well as a boom in the petroleum industry. The success of the telephone and wireless telegraph changed the way people transmitted information and greatly improved production efficiency. GDP per capita continued to rise rapidly in the 1900s, which was also the contribution of the "Second Industrial revolution."

The "Third Industrial Revolution" (1940–now) has involved a great leap forward in science and technology since the steam revolution and the electric power revolution, with the invention and application of computers, nuclear energy, the Internet, and biomedical engineering as the main breakthroughs, and also including a digital revolution in the fields of information, new energy, new materials, and biotechnology. The closer the integration of science and technology, the faster the transformation of technology into a productive force and the more profound the interaction between various disciplines. In response to the rapid change of complex emerging interdisciplinary knowledge, schools and colleges of electrical engineering and computer science have become the mainstream avenues of learning in the 20th century. The invention of computers was a major milestone in the "Third Industrial revolution." The development of computers was mainly due to World War II. Both ballistic analysis and nuclear fission research required efficient computing tools, thus laying a solid foundation for classical computers and accelerating the advent of the information age.

As quantum science matured, scientists managed to grasp microscopic laws and develop condensed matter physics to understand the properties of crystals formed by the periodic arrangement of atoms, which allowed us to differentiate between conductors, insulators, and semiconductors. This laid the foundation for the transistor industry. In 1947, William Shockley, John Bardeen, and Walter Brattain of Bell Labs invented the transistor. Transistor can effectively amplify signals and quickly switch between 0 and 1, performing binary operations. With the use of transistors, the instruction cycle of classical computers was accelerated, and the size and power consumption were also reduced. The "Very Large-Scale Integrated Circuit" (VLSI) and "Microprocessor" that followed the transistor further contributed to the advent of the personal computer era.

Due to the rapid development of the semiconductor manufacturing process, the number of transistors that can be accommodated on a chip doubles every 18 months. Within the same area, the number of transistors increased from less than 10 in the 1960s to 100,000 in the 1980s, and reached 10 million in the 1990s. Today, with the sophisticated technology of the semiconductor fabrication plant (fab), the number can easily reach hundreds of millions to billions. The exponential increase in the number of transistors on an integrated circuit is called Moore's Law. With the progress of Moore's Law, quantum tunnelling and nanotechnology are becoming increasingly important, and development bottlenecks have started to appear due to the limitation of the quantum size effect.

With the establishment of the Internet, various types of computers can be connected on the network, and data transfer is therefore more convenient. In addition, advanced chip design capabilities and the development of artificial intelligence (AI) have enabled direct communication between machines, opening up the era of Industry 4.0 or the Internet of Things (IoT), also known as the "Fourth Industrial Revolution" or "Productivity 4.0."

The key element of Industry 4.0 is a cyber-physical system (CPS), which is not only highly automated but can also actively remove obstacles to production in the era of the IoTs. The technical core of the IoTs is connectivity. Machines communicate and coordinate with each other by exchanging data, enabling more efficient automatic configuration of

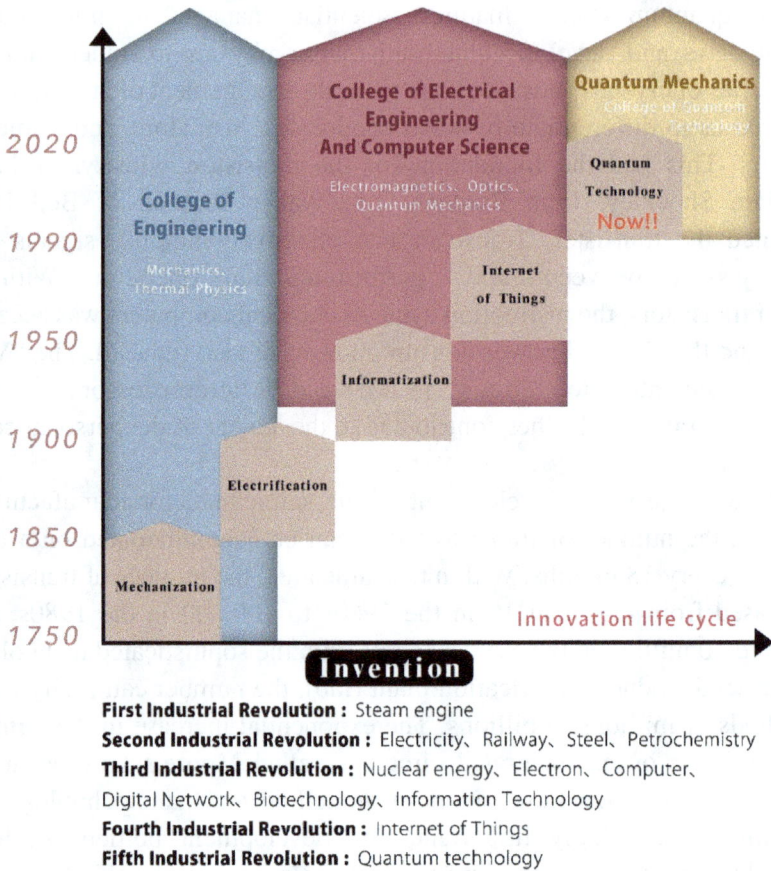

First Industrial Revolution : Steam engine
Second Industrial Revolution : Electricity, Railway, Steel, Petrochemistry
Third Industrial Revolution : Nuclear energy, Electron, Computer, Digital Network, Biotechnology, Information Technology
Fourth Industrial Revolution : Internet of Things
Fifth Industrial Revolution : Quantum technology

Figure 1.1.   Each Industrial Revolution has propelled human technology forward like waves, accompanied by the transformation of knowledge and the birth of new colleges. The gradual maturation of quantum technology will give rise to the establishment of quantum technology institutes, paving the way for a potential Fifth Industrial Revolution that may follow the era of Industry 4.0.

network resources, and building massive amounts of information into an intelligent digital world. Combined with the computing power of central supercomputers and edge computing of terminal sensors, instant messages on the network are collected at any time, and high-performance computing and artificial intelligence are used to make various complex analyses and real-time judgements. The current technological challenge is to monitor and ensure continuous quality services in the IoTs era. Analyzing massive amounts of data and optimizing resource allocation require high-performance computing, but the world's highest-performance supercomputers still cannot satisfy people's desire for computing power.

1.2 The background and characteristics of the second quantum revolution

1.2.1 *Where is the quantum valley?*

From the end of the 20th century to the beginning of the 21st century, Silicon Valley nurtured many technology giants and unicorns. At present, the world is actively looking for the quantum valley of the emerging quantum technology. Canada has invested in quantum technology for more than 20 years, and its technology and manpower are relatively mature. Therefore, Waterloo, Canada, has established a quantum valley to gather the world's resources and manpower. In April 2021, the Chicago Quantum Exchange was founded in Chicago to create the world's first quantum technology incubation hub, hoping to create the next Silicon Valley in the United States. The Quantum Technology Incubation Center mainly focuses on the three commercial fields of quantum computing, communication, and sensing, supporting quantum start-up companies in the development of industrial applications, such as building a "quantum Internet" that hackers cannot attack, developing quantum computers, and creating medical sensors that detect cell activity. Professor David Awschalom, Director of the Chicago Quantum Exchange, said that quantum technology is likely to be the next hot industry, with 1 million quantum engineer job openings expected in the United States over the next decade. AT&T partnered with Caltech to form the Alliance for Quantum Technologies (AQT), whose goal is to bring industry, government, and academia together to speed up the development and emerging practical applications of quantum technologies. The Netherlands is also building "Quantum Delta" in an attempt to become Europe's quantum valley. Recently, The Ministry of Economic Affairs and Climate Policy led the establishment of the House of Quantum, which is responsible for promoting the quantum industry cluster. In 2021, Germany's "Munich Quantum Valleys" was launched, focusing on developing quantum computing technology, secure communication methods, and basic research on quantum technology. "Quantum Avenue" has also appeared in Hefei, Anhui, China. Companies distributed along the "Quantum Avenue," such as Origin Quantum, QuantumCTek, and CIQTEK(Guoyi QuantumTechnology), are engaged in quantum computing, quantum communication, and quantum sensor-related industries, as well as building a quantum valley. Quantum technology has evolved from academic research to industrial production, but the general public still often asks when the era of quantum technology will begin. The answer is that quantum supremacy has arrived, and it is imperative to immediately believe in quantum, learn quantum, and use quantum.

1.2.2 *The first quantum revolution*

After Europe and the United States proposed the concept of the "Second Quantum Revolution," a natural question arises: "When and what were The First Quantum Revolution?" The current consensus is that technology that only utilizes quantum phenomena produced by finite size effects belongs to the category of the "First Quantum Revolution." In short, most of the science of nanotechnology belongs to The "First Quantum Revolution," such as the energy band of semiconductors and the tunnelling effect in nanostructures. The specific technological products developed during this time were CMOSs, computers, lasers, LEDs, solar cells, smartphones, the Internet, etc.

1.2.3 *The second quantum revolution*

Although the term quantum itself is esoteric and difficult to understand, it has now become a popular word and is often added as an adjective before existing words, causing many misunderstandings by the general public. Therefore, before determining the scope of the "Second Quantum

Revolution," it is necessary to understand what things are just riding the quantum boom to catch the eye.

(a) **In name, but not in fact:** These things have nothing to do with the properties of quantum mechanics. Words such as quantum investment and quantum leap have actually appeared in all levels of society, and are just adjectives that describe sudden changes in things. However, more strange usages have appeared recently, such as people meeting and greeting each other with "Are you quantum entangled?" There are also many commercial products with the word "quantum," such as quantum insole, quantum steamed bread, quantum rice, quantum speed reading, and quantum water, but the meaning of these is unclear. These phenomena can only represent society's misconceptions about quantum, and it is unclear if they will be of any specific help in the "Second Quantum Revolution." In fact, the marketing strategy of citing scientific and technological terms to attract consumers and catch the public's attention has existed for a long time. It takes advantage of the public's longing to know about advanced technologies by including buzzwords. For example, a ballpoint pen is translated as "原子筆" in the Eastern world, which means "atom pen." Regular blankets are sold in stores as "space blankets" (太空被) in Taiwan. Other examples are quantum water and AI glasses.

(b) **Old wine in new bottles:** There is indeed a quantum effect, but it is not what the "Second Quantum Revolution" emphasizes. There are mainly two quantum effects in this category: "quantum confinement effect" and "quantum tunnelling effect." In nanotechnology, as the sizes of objects get smaller and smaller, the finite size leads to the "quantum confinement effect," which is the principle of light emission of quantum dots. Therefore, TVs made with quantum dots have recently been referred to as "quantum TVs." Although this involves the quantum effect, it is purely old wine with a new label, trying to ride the "Second Quantum Revolution" to catch the attention of laymen and make a profit.

(c) **The focal points of the "Second Quantum Revolution":** The "Second Quantum Revolution" is defined by innovative applications that utilize quantum entanglement, quantum superposition, and quantum measurement. Currently, there are four major research and development directions as mentioned in the following:

1. *Quantum computers*: This involves providing a programmable universal computer that takes advantage of quantum superposition and entangled states in solving specific problems that classical computers cannot handle.

2. *Quantum computation*: This involves using quantum computers to simulate the dynamics of complex quantum many-body systems in nature, such as material properties and chemical reactions. It is also helpful in solving various problems encountered in fundamental science, material development, quantum chemistry, and industry. In addition, the optimization of massive amounts of data in society is also a key direction of quantum computation.

3. *Quantum communication*: This involves an anti-eavesdropping communication method by computing, encoding, and transmitting information with quantum no-cloning property to establish a secure communication network. The main technologies include Quantum Key Distribution (QKD), Quantum Teleportation, and Quantum Secure Direct Communication (QSDC).

4. *Quantum sensors*: This involves using quantum superposition and quantum entanglement to develop new quantum sensing devices. Since quantum states are extremely susceptible to changes in the external environment, the sensitivity and resolution of quantum sensors can significantly exceed the classical limit. Quantum sensing technology has a wide range of applications and can play a great role in the fields of acoustics, optics, mechanics, electronics, and thermodynamics.

After sufficient development of these four quantum technologies, they can be combined into a quantum IoTs, which is more sensitive, faster, smarter, and more efficient than the current IoTs. Quantum technology on the quantum IoTs can improve computing power and enhance real-time analysis and optimization speed, and the anti-eavesdropping quantum network is the best guarantee for the secure transmission of the quantum IoTs.

1.3 The global quantum technology race in the post-silicon valley era

The focus of the "Second Quantum Revolution" is on surpassing the limits of classical technology by applying quantum properties to traditional tools to improve their performance or by developing new tools and technologies. An important point in the "Second Quantum Revolution" is that nanoscale objects do not necessarily have quantum entanglement and other quantum properties. It is essential to utilize properties such as quantum superposition, quantum entanglement, and quantum measurement to design functional components of the "Second Quantum Revolution." These quantum properties in the components will not appear simply due to the continuous progress of miniaturization technology and Moore's Law. In fact, as long as the behaviour of a system is clearly understood, even sub-micron technology can make quantum components with quantum entanglement. Since the performance of components with quantum entanglement is far superior to that of classical electronic components, the quantum components and quantum computing methods of the "Second Quantum Revolution" will bring about disruptive industry innovation.

As can be seen from the technological trends in recent years, the "Second Quantum Revolution" is booming worldwide. Due to the rapid progress of fundamental scientific research and the gradual maturity of applied quantum technology, countries around the world are realizing that quantum technology will reshuffle the national power rankings, and the "Second Quantum Revolution" will precipitate the struggle for future world hegemony. As mentioned in a Forbes article in 2019, "We are witnessing a quantum battle that will be fought in research labs by brains instead of guns, and by scientists instead of soldiers. ...... If the United States loses, the long-term effect will be chilling." At present, this invisible technology world war is already in full swing and has even gone beyond the stage of scientists replacing soldiers. Instead, armies of millions of quantum engineers will face a head-to-head technology war.

According to statistics, by 2023, the total research and development funds invested in quantum technology worldwide will have reached 55.4 billion US dollars. In addition to China and the United States actively competing for the leading position in quantum technology, traditional Eurasian technology powerhouses are also trying to catch up. What is noteworthy is that some emerging countries are also vigorously participating, mainly because the leading group in quantum technology is not too far from the starting line. The chance of making swift progress and overtaking others is more likely in quantum technology than in semiconductor technology. Therefore, even BRICS countries are not falling behind with investment in quantum technology. For the convenience of the readers, we only discuss the recent development of China and the United States in the following (readers who are interested in the current situation in other countries and regions can refer to Appendix A):

(a) **United States:** Since the late 1990s, the US government has funded programmes to explore how quantum technology can assist US national security, and universities and private companies have also been actively investing in quantum information-related research and development. For example, the National Science Foundation (NSF) announced the Quantum Leap in the "Ten Big Ideas" in 2017, which uses quantum mechanics to develop new quantum technologies for sensing, computing, modelling, and communicating. Furthermore, the National Quantum Initiative Act (NQI) was passed in 2018, which invested more than 1.2 billion US dollars in accelerating the research and development of quantum technology within 5 years. The following year, the National Quantum Initiative Advisory Committee (NQIAC) was established to assess trends and developments in quantum information science and technology, as well as the progress of the NQI. In October 2019, Google demonstrated quantum supremacy with its 53-qubit quantum processor "Sycamore." The Google team claimed that Sycamore only took 200 seconds to finish a task that would take 10,000 years for the world's best supercomputer to complete. In February 2020, the White House National Quantum Coordination Office (NQCO) released "A Strategic Vision for America's Quantum

Networks," proposing that the United States build a quantum Internet so that the American people can enjoy the convenience brought about by quantum technology. Later that year, the US Department of Energy unveiled a report that set a strategic blueprint for establishing a national quantum Internet. In 2021, US President Joe Biden said US 180 billion would be committed to "R&D and industries of the future," such as quantum computers and semiconductors. In response to technological competition with China, the US Senate and House of Representatives proposed the "Endless Frontier Act" in the same year. In addition to the US 112 billion for science and technological research, the bill also plans to invest US 10 billion in at least ten regional technology hubs to maintain US leadership in semiconductors and quantum technology. In May 2022, President Biden further signed the "National Security Memorandum" (NSM), instructing the United States to maintain world dominance in quantum information science. Both IBM and Honeywell announced roadmaps for quantum computers in early 2021, expecting practical quantum computers with more than 1,000 qubits to be available in 2023 and 2028, respectively. IBM further announced in 2022 that a 4,000-qubit commercial quantum computer is being developed.

(b) **China:** China is one of the leading countries in quantum communication. In December 2011, it officially launched a quantum communication research and development project. In 2016, China successfully launched the world's first quantum science experimental satellite "Micius" into space. The primary purpose of the satellite was to conduct experiments on quantum key distribution between space and the ground. In 2018, satellite-to-earth quantum key distribution was demonstrated by Micius over 7,600 km between Xinglong in Hebei, China, and Graz in Austria. Furthermore, in 2022, quantum state transfer over 1,200 km was demonstrated between Lijiang in Yunnan and Delingha in Qinghai, which is an important step to build a global quantum information processing and communication network. The world's first quantum micro–nano satellite "Jinan-1" has also been successfully operated since 2022. In 2022, a team led by Professor Gui-lu Long from Tsinghua University designed a new quantum direct communication system with mixed encoding of phase quantum state and time-bin quantum state, achieving 100-km quantum secure direct communication. This breaks the world record of quantum secure direct communication.

On October 16, 2020, the Political Bureau of the Communist Party of China's Central Committee conducted a study on quantum technology, emphasizing the importance and urgency of promoting the development of quantum technology. During the study session, General Secretary Xi Jinping stressed that China would strengthen its strategic planning for the development of quantum technology. In December 2020, the University of Science and Technology of China (USTC) announced that it had success-fully built a prototype quantum computer "Jiuzhang" of 76 photons, making China the second country to achieve quantum supremacy. The development team claimed that it takes only 200 seconds for Jiuzhang to solve the Gaussian boson sampling problem of 50 million samples, com-pared with the supercomputer Fugaku, which required 600 million years; when solving the problem of 10 billion samples, Jiuzhang takes 10 h, while Fugaku needs 120 billion years. In June 2021, USTC devised a programmable 56-qubit superconducting quantum computer, "Zuchongzhi," which shortened the computational time of a task that takes 8 years for a supercomputer to complete to 1.2 h, demonstrating the tremendous com-putational advantage of quantum computing. At the end of 2021, USTC further developed "Zuchongzhi 2" and "Jiuzhang 2," which have stronger computing power, with both physical systems reaching the milestone of "quantum computational advantage." This has led European quantum experts to assess that China's quantum computing technology may already be on par with that of the United States. The "14th Five-Year Plan" of China, starting from 2021, calls for accelerating the development of quan-tum computing and quantum communication, with the goal of building a national quantum communication infrastructure and developing a univer-sal quantum computer by 2030. The Chinese government is estimated to have invested at least US 15 billion in quantum technology. Not to be outdone by IBM and Honeywell, Origin Quantum announced its quantum computer roadmap in September 2021, aiming to launch a 1,024-qubit quantum computer in 2025. There is a "Quantum Avenue" in the Hefei

National High-tech Industry Development Zone, where China's core quantum companies, such as QuantumCTek, CIQTEK, and Origin Quantum, are located. The upstream, midstream, and downstream industries of quantum communication, quantum computing, and quantum precision measuring instruments are clustered around about 100 m of road in this area. At present, there are more than 40 quantum technology companies at the Hefei High-tech Zone, in which many of the talents are from USTC. In 2022, "Quantum Science Instrument Valley" was established in the Hefei High-tech Zone. The Hefei High-tech Zone aims to build a world quantum centre with the concept of "one core, four parks, and one city": "One core" refers to the "innovation source core" centred on the national laboratory, "four parks" refer to the four major industrial clusters represented by the Quantum Science Instrument Valley, and "one city" refers to "Quantum Future Technology city."

1.4 Classification and industry status of quantum computer

1.4.1 *What is a quantum computer?*

To put it simply, quantum computing is a combination of quantum mechanics, linear algebra, and the theory of computation, which exploits quantum superposition, quantum entanglement, and quantum interference to obtain computational power beyond classical computers. Currently, quantum computers can be divided into four types:

1. **Universal quantum computer:** This is a computer that harnesses the properties of superposition and entanglement states and performs programmable operations based on quantum logic gates.
2. **Special-purpose quantum computer:** This is divided into two categories: quantum annealer and analog quantum simulator. Quantum annealers utilize quantum properties of quantum components to solve NP-hard problems, such as specific optimization problems and graph theory problems. Analog quantum simulators are artificial quantum systems that can be manipulated in a controlled way, designed to simulate quantum systems in nature.

3. **Quantum-inspired computer:** This type of computer uses electronic components to simulate the quantum tunnelling effect and specializes in solving optimization problems.
4. **Educational quantum computer:** Due to the very limited number of qubits, such computers can only be used for teaching purposes.

Sensitive quantum states are easily destroyed by external thermal disturbance and electromagnetic interference. Currently, the main method is to keep the quantum system as isolated as possible and place qubits in an environment close to absolute zero to prolong the coherence time, maintaining superposition and entanglement property of state. The leading qubit fabrication technologies are superconductors, ion traps, cold atoms, nano-diamonds with nitrogen vacancy, quantum dots, silicon-based qubits, topological qubits, photonic integrated circuits, nuclear magnetic resonance (NMR), etc. Based on the physical systems they utilize, these technologies are summarized in Figure 1.2. Among them, superconductors, ion traps, and photonic integrated circuits are technologies that are currently close to commercialization. At present, major companies in the

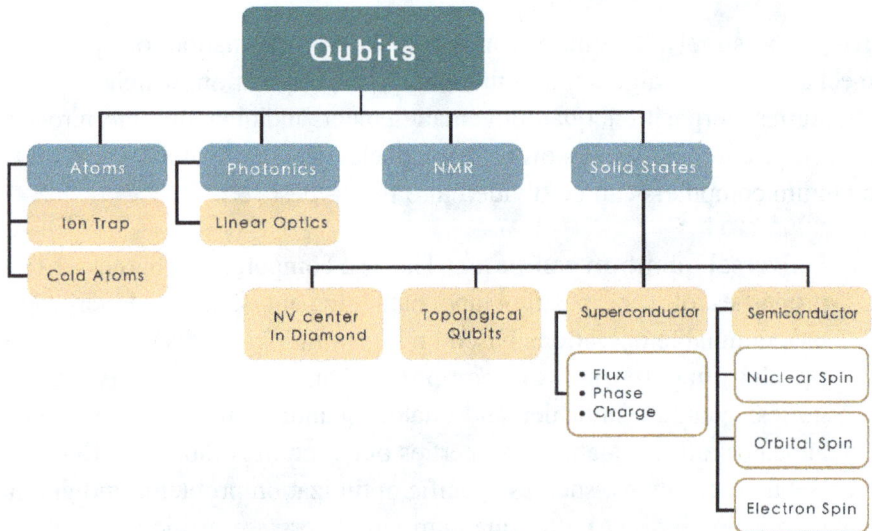

Figure 1.2.   The current leading fabrication technologies of physical qubits.

world are mainly focusing on developing universal quantum computers. The progress of these companies can be found in Appendix B.

1.4.2 *Moore's Law for classical computers, Rose's Law, and Neven's Law for quantum computers*

The exponential growth in the power of digital computers since 1949 is known as Moore's Law. In 1965, Intel co-founder Gordon Earle Moore noticed that the number of transistors on an integrated circuit was regularly doubling. In 1975, Moore predicted that computing power would double every 2 years. D-Wave co-founder Geordie Rose proposed Rose's law for quantum computers. His main observation was that the number of qubits has doubled every 5 years since 1989. Besides the exponential growth in the number of qubits, the dimension of computing space formed by entangled qubits also exponentially increases with the number of qubits. In 2018, Google scientist Hartmut Neven proposed Neven's Law of double exponential growth. The exponential growth brought about by improvements in hardware manufacturing processes and qubit entanglement combine to produce the double exponential Neven's Law for quantum computers, which is more amazing than Moore's Law for classical computers.

1.5 Concluding remarks

The Industrial Revolution in the history of modern science proceeded gradually, from the internal combustion engine that utilized machinery, force, and heat to the petrochemical industry, electronic industry, and the Internet, one after another. It is only a matter of time before quantum technology matures. Google, IBM, Honeywell, and Origin Quantum have recently given mass production schedules for quantum computers. In principle, there will be a fault-tolerant universal quantum computer around 2030 that is able to solve problems in biomedicine, agriculture, and finance. Therefore, some people state that 2030 will be the "Year to Quantum" (Y2Q). IBM is even more optimistic that 2029 will be a turning point for quantum error correction. Google CEO Sundar Pichai said,

"Quantum computing represents a fundamental change, as it takes advantage of properties of quantum mechanics, it gives us the best opportunity to understand the natural world." Consulting firm Gartner noted that inquiries about quantum computing by technology executives of various enterprises have increased by 28% in recent years. Gartner predicts that 40% of the world's large enterprises will set up quantum computing departments by 2025, and quantum computing will become the future trend of the technology sector. In fact, companies such as Visa, JPMorgan Chase, and Fox have all begun to use quantum computing to analyze massive amounts of data and have achieved considerable success. The business community hopes to use the computing power of quantum computers to develop renewable energy, reduce carbon emissions, and achieve sustainability to support the rapidly increasing population on earth.

Since the Industrial Revolution, the change in the world's technology has always been dramatic. The knowledge applied in engineering colleges is mainly classical physics, such as mechanics and thermodynamics, while electrical engineering colleges apply the knowledge of optics and electricity, which is conventional in the 20th century. What other knowledge will be transformed into an industry in the 21st century? It is obvious that quantum technology has emerged. Many countries have started setting up colleges of quantum technology to recruit students. The situation is similar to the emergence of electrical engineering 60 years ago. Quantum technology will become a global trend sooner or later. Quantum technology is an interdisciplinary technology, spanning at least system engineering, materials science, cryogenics, software engineering, semiconductor technology, and photonics. Moreover, software applications of quantum technology require the integration of all fundamental and application domains to have a significant impact. Many countries have begun to prepare for the advent of the quantum era, setting up departments or colleges of quantum science and technology to reserve talents of quantum technology. The promotion of quantum technology not only requires the joint efforts of industry, government, and academia but also requires the integration of the K-12 education system. The collaboration between upstream, midstream, and downstream industries is necessary to push quantum technology forward; there is no way a single company or specialist in a single field can do it independently.

Due to historical reasons, China has missed the opportunities of the previous technological revolutions since the Industrial Revolution, resulting in a passive situation where it is now struggling to catch up. Professor Jian-Wei Pan of USTC, who has been called the "Father of Quantum" by *Nature* magazine, said, "In modern information science, China has always played the role of a learner and a follower. Now in the era of quantum technology, if we do our best, we have a chance to be a leader." Professor Pan has at least two achievements in quantum communication that surpass the Western world. The first is the "Micius" satellite for ground–space quantum communication in 2016. The second is the 2,000-km quantum communication line between Beijing and Shanghai that was set up in 2017. These two achievements caused the Pentagon's to report to the Congress in 2021 that "China seeks leadership in key technologies with significant military potential." In 2021, the United States even directly included QuantumCTek on the export control list.

There are four major directions for quantum technology at present: quantum computers, quantum computing, quantum communication, and quantum precision measurement. China and the United States are currently rivalling each other in the development of quantum technology. In the field of quantum computers and quantum computing, IBM and Google in the United States still have technical advantages. According to the quantum computer roadmaps released by Origin Quantum and IBM in 2022, there is at least a gap of about 3 years between the two countries. On the other hand, in the field of quantum communication, China is in a leading position in terms of basic research, network construction, and even the influence of patents and articles. When it comes to the field of quantum precision measurement, China's scientific research level and technical applications are on par with those of European countries and the United States. According to "Quantum Technologies: A Review of the Patent Landscape" published by Relecura Technologies Pvt. Ltd., among the 44,394 patents in quantum technology filed for review from 2015 to 2020, China leads the world in the total number of patents. Except for quantum computers and quantum computing, China holds several times more patents than the United States in every other category. Compared with previous technological revolutions, one of the main reasons why China's quantum technology can keep pace with the world is that China's scientific and technological research has indeed taken off. The other reason is that China has the institutional advantage of concentrating resources on important issues. Up to now, commercial products of quantum technology have not been available in a short period of time. Still, quantum technology is a key technical field related to future national power, and there is an urgent need for more young people with both intelligence and courage to break into the no-man's-land of scientific research and industry, becoming forerunners of the quantum era. The current shortage of quantum talent is global. The United States advocates including courses on quantum science in education from kindergarten to high school, and Harvard University has launched a quantum science and engineering Ph.D. programme.

During the 24th joint study of the Political Bureau of the Communist Party of China's Central Committee, Xi Jinping stressed that "the world today is undergoing great changes unseen in a century, and China must seek answers from technological innovation. Therefore, it is necessary to recognize the importance and urgency of promoting quantum technology's development and strengthen its strategic planning." To speed up the training of talents in the field of quantum science and technology, it is crucial to establish a special training plan suitable for quantum technology development and create a systematic and high-level training platform. In China, USTC has been approved to set up an undergraduate major in quantum information science and a Ph.D. programme in quantum science and technology. Tsinghua University has also established an undergraduate programme in quantum information. At the beginning of 2021, Professor Shing-Tung Yau of Tsinghua University ambitiously declared, "In ancient times, Huo Qubing fought against Xiongnu in the north, marched over thousands of miles and defeated the enemy in Mobei (Northern Desert) with eight hundred cavalries. I hope that through this plan, we can cultivate our own eight hundred cavalries in fundamental science."

Quantum technology is a future technology of irreplaceable strategic importance, so it is bound to be the focus of China–US technological competition. In 2018, the American think tank "New America" proposed a defence strategy of "small yard, high fence" regarding core technologies, first identifying scientific research areas directly related to US national security ("small yard") and delineating appropriate strategic boundaries ("high fence"). According to the strategy, the core technologies within the "small yard" should be strictly protected, while the high-tech fields outside the "small yard" should be opened up. In December 2018, US President Donald Trump signed the National Quantum Initiative Act, raising quantum technology to a national strategic level. In August 2020, the US government announced about US 1 billion in funding for quantum information science (QIS) and artificial intelligence, including 925 million for QIS and 140 million for AI. According to Paul Dabbar, Former Undersecretary of Energy for Science, quantum science may have a bigger impact on US national interests than artificial intelligence. Since Biden took office, he has basically implemented a strategy of "asymmetric competition" against China, that is, the "bifurcation" strategy of selective decoupling from China and the "small yard, high fence" strategy. The United States is also preparing to form a new technology alliance with US allies, including Japan, Germany, France, the United Kingdom, Canada, the Netherlands, South Korea, Finland, Sweden, India, Israel, and Australia. This new alliance, currently called the "T-12" Forum, is being urged by the United States to jointly respond to technological competition from China. If the confrontation between China and the United States intensifies, the T-12 will likely move towards the technology blockade strategy of the technology island chain.

The quantum computer race began with Google's Sycamore, and China subsequently demonstrated that the photonic quantum computer "Jiuzhang" has a greater advantage in solving specific problems. Furthermore, Xanadu recently launched "Borealis," a fully programmable photonic quantum computer that is far faster than the world's fastest supercomputer. We have reached a historical turning point in the development of quantum computing, and quantum technology will begin to show its power on the international stage. The understanding of the microscopic world will be clearer, and the technology services in the macroscopic world will be better. However, the world's technological competition may also intensify in the short term.

There are various signs that the "Second Quantum Revolution" has started, and Y2Q is not far away. Those over 55 years of age may not need to learn much about quantum technology because most of them will be retired soon, but the young "quantum generation" will not be able to escape the upcoming quantum technology ecosystem. Those over 55 years of age still need to encourage and support the next generation to invest in quantum technology; otherwise, every country's global competitiveness will gradually disappear due to the worldwide rise of quantum technology. The rapid development of quantum technology around the world is expected to have a significant impact in the next few years. The leading countries in the semiconductor industry have a good foundation for developing quantum technology, but it does not guarantee that they will have definite advantages in the development of quantum technology. This is because the "Second Quantum Revolution" is distinct from the current industry in terms of principles, materials, cryogenic technology, and peripheral devices. History keeps telling us that when disruptive and revolutionary technologies appear, the original advantages can turn out to have fatal effects because of people's tendency to follow what has been established. The Icarus Paradox provides those who have not caught the rapidly progressing "Semiconductor Technology Revolution" train with more room for imagination to change lanes and overtake in the quantum technology revolution.

China's current technological power is vastly different from the early years of semiconductor technology development, when the whole nation was focused on the pursuit of a stronger national defence in an attempt to fill the gap from the discoveries of quantum science to the Manhattan Project era. China established the independence of its defence industry at the time, but it missed the opportunity to ride the booming semiconductor industry wave, due to which it is now struggling to catch up on chip technology. Nevertheless, China's technological power has gradually become more and more impressive in recent years. Tsinghua University has gathered three "first Chinese winners of very prestigious prizes": the first Chinese Nobel Prize winner, Professor Chen-Ning Yang, the first Chinese Turing Award winner, Professor Chi-Chih Yao, and the first Chinese Fields Medal winner, Professor Shing-Tung Yau. The performance of the younger generation in technology is not inferior, especially in the emerging quantum technology, in which China has entered the world's leading group. Quantum technology is by no means a short-term competition between countries but a long-term collective competition between humans and nature. In the context of an intense global rivalry, it is the responsibility of all contemporary Chinese people to build a technological cavalry, as Mr. Shing-Tung Yau stated, to break through the global technology blockade and recreate the prosperous Han and Tang Dynasties in the quantum era. In the post-Silicon Valley era, how to mobilize the entire nation's technological strength in the fireless war of the "Second Quantum Revolution" is an inescapable historical responsibility and destiny for this generation of Chinese people, as they prepare to decisively battle in the Quantum Valley.

Quantum supremacy has arrived, and quantum superposition and entanglement have been actively utilized to develop technology around the world. The "Second Quantum Revolution" will determine where the future hegemony lies. How the situation will evolve remains to be seen.

Chapter 2

History of Quantum Theory and Related Concepts

When you haven't looked at this flower, this flower and your heart go to silence together; when you look at this flower, the colour of this flower becomes clear for a moment.

— Wang Yangming (1472–1529)

Does the moon only exist when you look at it?

— Albert Einstein (1879–1955)

2.1 The origin and background of quantum mechanics

2.1.1 *The conference that changed human history*

The fifth Solvay Conference (Conseils Solvay) was held in October 1927 on the subject of "Electrons and Photons." This physics conference was an unprecedented showdown between the best minds in history. Seventeen of the twenty-nine people invited that year to the conference won the Nobel Prize. The group photo of the 29 participants (Figure 2.1) is often referred to as the "the most intelligent photo ever taken." It was the most intellectually influential meeting ever, and quantum theory has never been in doubt since. The collective minds of the participants created the scientific

Figure 2.1.   Participants of the Fifth Solvay Conference.

foundation of the modern world and started the quantum revolution that continues to this day. There were three camps at the meeting. The Copenhagen school led by Bohr supported the probabilistic interpretation of quantum mechanics; the opposition led by Einstein insisted on determinism, and Bragg and Compton emphasized experimental results. There were also neutral spectators such as Marie Curie. In 1926, Einstein wrote to Max Born to oppose Born's probability interpretation with the view that "God does not play dice." In this meeting, he also put forward the same argument to question quantum probability, while Bohr refuted Einstein with "It cannot be for us to tell God, how he is to run the world." This dialogue has been widely interpreted as a classic statement of Einstein's skepticism toward quantum theory. There was a heated debate about how to interpret Schrödinger's wave equation during the conference. Heisenberg and Born finally came to the conclusion that quantum mechanics is already complete and no longer needs more modifications, officially declaring the quantum revolution a success. However, after the meeting, everyone still held their own opinions. The new thinking generated by the fierce debates actually accelerated the progress of quantum science. At the sixth Solvay Conference 3 years later, although the subject of the conference was "Magnetism," Einstein once again proposed a "photon box" thought experiment to challenge the core concept of quantum theory — the uncertainty principle. Einstein considered a box with a quick shutter, which lets exactly one photon escape at a time. Since the energy of the photon can be determined by measuring the weight of the box before and after the escape of the photon, it appears to prove that the energy and time of photons can be accurately determined simultaneously. Bohr was unable to respond immediately on the spot, but he kept trying to convince other participants that Einstein could not be right, otherwise physics would be over. Einstein was very proud of his argument all day. After staying up all night thinking hard, Bohr finally came up with an excellent way to fight back against Einstein. The next morning, Bohr explained using general relativity that after the photon is emitted, the clock that controls the shutter would shift along the direction of the gravitation field because the box becomes lighter, so the clock's ticking rate would change. Therefore, energy and time still could not be precisely measured simultaneously, and he argued that the uncertainty principle remained correct in the "photon box," refuting Einstein with his own theory. Bohr triumphantly said that physics was finally saved, and Einstein's challenge failed again. Afterwards, due to the Nazis and World War II, Einstein did not attend the seventh Solvay Conference. But, he still published the famous Einstein–Podolsky–Rosen (EPR) paradox in 1935, again designing a thought experiment to emphasize the incompleteness of quantum mechanics through the long-distance correlation exhibited by two entangled particles, trying to make a comeback with the so-called "spooky action at a distance." Despite Einstein's persistent resistance to quantum theory , he could not hold back the overwhelming quantum wave. In his later years, when walking with his colleague Abraham Pais, Einstein was still wondering and asked Pais, "Does the moon only exist when you look at it?"

After the dust settled on the dispute, quantum mechanics immediately showed great power. The United States first applied quantum theory and relativity in the Manhattan Project led by Oppenheimer, invented the atomic bomb, and quickly ended World War II. After the war, the

United States developed the classical computer based on CMOS technology, taking the lead in the world's technological race. The advent of the CMOS computer also greatly promoted technological development across the world. With the rapid development of quantum technology products, such as photonic and semiconductor devices, the world's population and economy have continued to grow steadily. At the end of 2018, the European Union held the Quantum Flagship Kick-Off Conference in Vienna, claiming that although quantum science originated in Europe, it is the United States that has long enjoyed the fruits of quantum technology. To let European civilization shine again, Europe not only launched the Quantum Flagship initiative but also formed a European quantum fleet for the "Second Quantum Revolution." To understand why the quantum concept has such a powerful and world-changing influence, we must look back at the development of quantum theory.

2.1.2 *Dark clouds are forming, heavy rain brings new opportunities to the world*

At the end of the 19th century, physicists felt that they already had a clear understanding of the laws of nature. The development of physics was quite complete. Based on Newton's classical physics and Maxwell's electromagnetic theory, supplemented by statistical thermodynamics and wave optics, various macroscopic phenomena could be accurately analyzed and predicted. The world economy has also benefited from the Industrial Revolution prompted by mechanics and thermodynamics. In addition to their great satisfaction, physicists gradually felt that the knowledge of physics had matured, and it seemed that there was no new opportunity and hope in the field of classical physics.

At the beginning of the 20th century, famous European scientists gathered at the Royal Institution in London. Lord Kelvin, the great physicist who formulated the second law of thermodynamics, gave a speech entitled "The Nineteenth Century Clouds over the Dynamical Theory of Heat and Light" at the meeting, pointing out that the beauty of the dynamical theory was overshadowed by two "dark clouds." One of the dark clouds refers to the relative motion between matter and luminiferous aether, which is a postulated medium pervading the space and thought to be the medium required for the propagation of light. The aether theories were not consistent with the results of famous Michelson–Morley experiment, which detected no relative motion between earth and aether. The other dark cloud refers to the equipartition theorem in thermodynamics, which deviates from the experimental results in explaining the thermal radiation spectrum. In particular, the discrepancy in black-body radiation is confusing. Classical physics cannot adequately explain the results of black-body radiation experiments. The first dark cloud gave birth to the theory of special relativity, and the second dark cloud gave birth to the quantum theory. In this visionary speech, Lord Kelvin made it clear that the development of classical physics has reached its limit, and also pointed out new possible directions. The new direction proposed by Lord Kelvin was like a beacon, illuminating the previously unknown microscopic world, and setting off an earth-shaking quantum revolution in the 20th century. Lord Kelvin never imagined that the dark cloud he mentioned would actually pour down torrential rain, nourishing the arid land of classical physics. Various offshoots of quantum physics sprang up one after another, forming the quantum era today.

2.1.3 *What is black-body radiation?*

As early as the Bronze Age, when humans smelted metals, they knew that high-temperature molten metals would emit different colours of light at different temperatures. This is the phenomenon of black-body radiation. The phenomenon of black-body radiation states that any object will emit electromagnetic radiation as long as its temperature is not absolute zero. Objects with different temperatures have different radiation spectra. Room-temperature objects also emit electromagnetic waves; it is just that the human eye cannot perceive the radiated infrared. The wavelength spectrum of black-body radiation is continuous. There is a specific peak in the spectrum. The higher the object's temperature, the more the peak position moves to shorter wavelengths (Figure 2.2 (a)). Therefore, the temperature of the object can be known by detecting the peak position of its spectrum. Recently, during the COVID-19 pandemic, non-contact

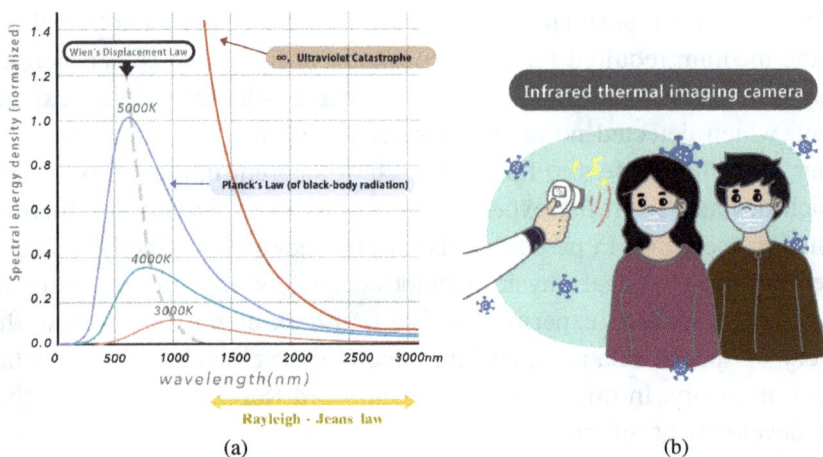

Figure 2.2. (a) Objects with different temperatures have different thermal radiation spectra, that is, energy density distribution curves. (b) A forehead thermometer measures body temperature by detecting the infrared radiation emitted by the human body.

forehead thermometers and thermal imaging cameras were being used to measure body temperature all over the world. The principle is exactly to use infrared sensors to detect the peak position in the emission spectrum of the human body. This method of measuring temperature is widely used, ranging from boiler temperature control to forehead thermometers. The biggest advantage of using black-body radiation to measure temperature is that it can be measured at a long distance. The measurement result does not depend on distance since only the peak position of the radiated spectrum is required to infer temperature. Distance only affects the sensitivity, not the peak position.

After the second Industrial Revolution, Europe vigorously developed the steel industry. But, when making steel, if you put a thermometer into the steelmaking furnace, it will melt immediately because of the ultrahigh temperature. So, then, how do we know the furnace's temperature? The black body is an ideal concept. Incident electromagnetic waves will be 100% absorbed by a black body. After a black body reaches thermal equilibrium with its environment, it will radiate electromagnetic waves. In the 1890s, German physicist Wilhelm Wien researched the black body's thermal radiation. He found that the peak wavelength of thermal radiation is inversely proportional to temperature, which is called Wien's displacement law. Simply put, there is a fixed relationship between the temperature of an object and the radiated energy, which also provides a new way to measure high temperatures. As long as the radiation coming out of the small hole of the steelmaking furnace is measured, the furnace's temperature can be determined according to the thermal radiation spectrum, that is, the shape of the energy density distribution curve. But, why does blackbody radiation have such a spectrum? Physicists at the time did not understand. Starting from classical statistical mechanics, the Rayleigh–Jeans law was derived to describe the spectrum of black-body radiation based on the equipartition theorem. The Rayleigh–Jeans law is consistent with the experimental results in the long-wavelength (low-frequency) part of the spectrum, but diverges to infinity at short wavelengths (high frequency). This divergence indicates that any object will radiate intense ultraviolet light at any temperature. This theory is obviously inconsistent with the experimental facts, which is the famous "ultraviolet catastrophe" in the history of physics. On the other hand, Wien's displacement law is in good agreement with experimental data in the short-wavelength range, but cannot explain the spectrum in the long-wavelength range. There is no unified explanation for the phenomenon of black-body radiation, which is the second dark cloud that Lord Kelvin talked about.

2.1.4 *Quantum theory popped out of nowhere!*

The problem with black-body radiation is that there are two formulas: One is valid only for short wavelengths of radiation spectrum and the other is valid only for long wavelengths. Moreover, Wien's displacement law is an empirical formula. Is there a fundamental law that describes the spectrum at both long and short waves? In 1900, in order to solve the problem of the "ultraviolet catastrophe," Max Planck published the celebrated Planck's law of black-body radiation. Just by introducing a strange mathematical parameter for the quantization of energy, Planck was able to combine the two previously known formulas perfectly. Planck's basic assumption is that there are countless small springs in the black body, and the energy of these small springs has a minimum basic unit, which is a "quantum." All the little springs can only move in states in which the energy is an integer multiple of the minimum energy unit. Each small spring is the source of electromagnetic radiation. If it absorbs radiation, the vibration accelerates; if it emits radiation, the vibration decelerates. The frequency at which the small spring vibrates is the frequency of the radiated electromagnetic radiation.

Planck introduced the historic Planck constant, which is the proportionality constant of the energy of a small spring to its frequency. Planck's law of black-body radiation perfectly fits the observed spectrum. Before Planck, no one dared to propose that energy is discontinuous, because this completely violates the basic understanding of energy in classical physics. In fact, Planck himself first thought it was just a mathematical trick and did not dare challenge the continuous concept of classical physics. Planck tried every means to explain the discontinuity using classical physics, but was unsuccessful. Because the quantum concept worked out so well, he had no choice but to embrace the beautiful mathematical result. Planck never expected his bold idea of energy quantum to cause earth-shattering changes in the field of science in the 20th century. The Planck constant is like a quantum master key, opening up discontinuous quantum new thinking and leading physicists into the quantum world. Planck's quantum theory of black-body radiation implies that there may be a minimum basic unit of any change. However, from the perspective of classical physics, all processes should be continuous. If the burning of a candle releases 10 J of energy, it must be a continuous change in released energy between the time it starts burning and the time it releases 10 J. It is impossible to jump from 5 J to 6 J or suddenly change to 10 J in a discontinuous manner. But, Planck's theory tells us that the change in energy is similar to how we spend money in our daily lives. If you go to a supermarket to buy a quarter-dollar drink with a 50% discount, you will either pay 12 cents or 13 cents at the checkout because the smallest currency denomination is 1 cent in the US. There is also an indivisible smallest unit for the energy in the quantum world; thus, the energy changes discontinuously. The concept of continuous and discrete change can be illustrated by an analogy of stairs and accessible ramps. Stairs are discrete, and the height at which one can stop is limited to an integer multiple of the step height, while ramps are continuous and one can stay at any height in the process.

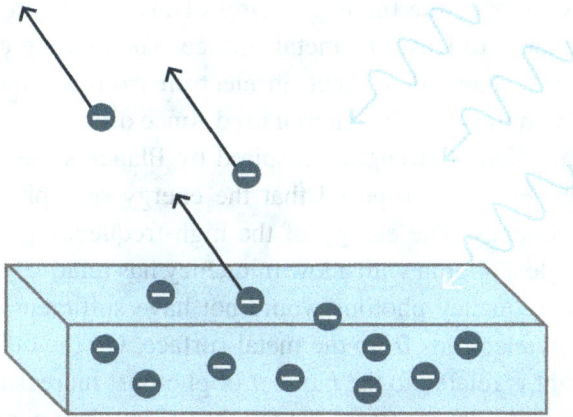

Figure 2.3. When photons collide with electrons in a metal, if the energy of the photon exceeds the binding energy of the electron to the metal, the electrons will acquire enough energy to escape from the surface of the metal.

Planck's quantum theory of black-body radiation fired the gunshot of the quantum revolution. Einstein, who was working at the Swiss Patent Office in Bern at the time, immediately used a similar idea to solve the photoelectric effect problem that had puzzled scientists for a long time. In the photoelectric effect experiment (Figure 2.3), when light is irradiated on a metal, electrons are ejected from the surface of the metal. What scientists could not understand was that whether incident light ejects electrons from a metal surface or not does not depend on the intensity of the light, but only on the frequency of the light. Regardless of light's intensity, high-frequency ultraviolet irradiation is always able to eject electrons, while a low-frequency light source, such as red or yellow light, is unable to eject even a single electron. This result is inconsistent with the electromagnetic wave theory which states that the radiant energy carried by light is proportional to the intensity. In 1905, Einstein successfully explained the photoelectric effect experiment using the quantum concept, and proposed the idea of "light quanta," now called photon, according to which light has both wave and particle properties. Einstein's idea of "photon" perfectly solves phenomena that classical electromagnetic theory cannot explain. Since each electron only interacts with one photon, only when the photon energy exceeds the binding energy of the metal can the electron gain enough energy to leave the metal surface. The binding energy is the minimum energy required to eject an electron from the metal surface, which depends on the specific material used, since different materials bind electrons with different strengths. Inspired by Planck's theory of black-body radiation, Einstein proposed that the energy of a photon depends only on its frequency. The energy of the high-frequency photon is relatively high, while a photon with a low frequency has relatively low energy. Since the low-frequency photons would not have sufficient energy, they cannot eject any electrons from the metal surface. On the other hand, the intensity of light is related to the number of photons. Increasing the intensity of the incident light increases the number of incident photons, but the frequency of the photons remains unchanged. As long as the frequency of the light source is lower than a certain cutoff frequency, it will not eject any electrons no matter how high its intensity is. Einstein's formula for the photoelectric effect also states that the kinetic energy of the ejected electron plus the binding energy of the metal is the energy of the incident photon, which is proportional to the incident light frequency according to his proposal.

Einstein's success in explaining the photoelectric effect using Planck's quantum concepts clearly overturned Maxwell's electromagnetic wave theory. Since Maxwell's electromagnetic theory was very successful in describing classical physical phenomena, many scientists, including Planck, did not accept Einstein's explanation of the photoelectric effect. When Planck recommended Einstein as a member of the Prussian Academy of Sciences in 1913, he still had reservations about quantum theory. He apologized for Einstein by saying, "That sometimes, as for instance in his hypothesis on light quanta, he may have gone overboard in his speculations should not be held against him." In 1916, Robert Andrews Millikan measured the magnitude of the current generated by the photoelectric effect under different voltage conditions using six different frequencies of light source. After analyzing the data according to Einstein's formula, a very accurate value of the Planck constant was obtained by Millikan. Millikan's experiment verified Einstein's theory of the photoelectric effect, and Planck and other physicists started to accept the idea of the photon. The photoelectric effect demonstrates that light possesses both wave and particle properties, laying an important foundation for quantum mechanics and wave–particle duality. Einstein inadvertently opened the quantum door with Planck's quantum key, leading countless scientists in the 20th century to the new quantum playground. Einstein received the Nobel Prize in 1921 for his explanation of the photoelectric effect, and 2 years later, Millikan also won the Nobel Prize for his accurate measurement of the charge of electrons and his work on the photoelectric effect.

2.1.5 *Spectrum is the fingerprint of matter*

Spectrum is like a human fingerprint, with each element having a unique spectrum. The composition of elements in an object can be determined by the colour of its flame, known as the flame test. The colourful patterns of the New Year's Eve fireworks in the sky are due to the coloured flames produced by burning certain compounds. For example, the flame colour is bright yellow for sodium salts, purple for potassium salts, red for lithium, and green for copper. The spectral lines can be obtained by passing these rays through a triangular prism and projecting them onto a screen. Each element has its own characteristic spectrum, and spectroscopy was therefore developed as a method for identifying chemical composition. By the mid-19th century, due to the development of spectroscopy and the advancement of measurement techniques, it was found that in addition to the black-body radiation spectrum, different hot rarefied gases also emitted spectral lines of different wavelengths (Figure 2.4). Among them, the spectrum of the hydrogen atom has attracted many scientists to study it in depth because hydrogen is the simplest atom, consisting of one electron and one proton. Scientists have learned more about the mysteries of quantum mechanics by studying the hydrogen spectrum.

In 1885, J. J. Balmer, a Swiss mathematics teacher, based on the measurement results of the hydrogen spectrum, found a formula that summarized a common rule for the wavelengths of the four spectral lines in the visible light region. Johannes Rydberg generalized Balmer's formula 3 years later, making it applicable to all hydrogen spectral lines. The Rydberg formula is expressed as follows:

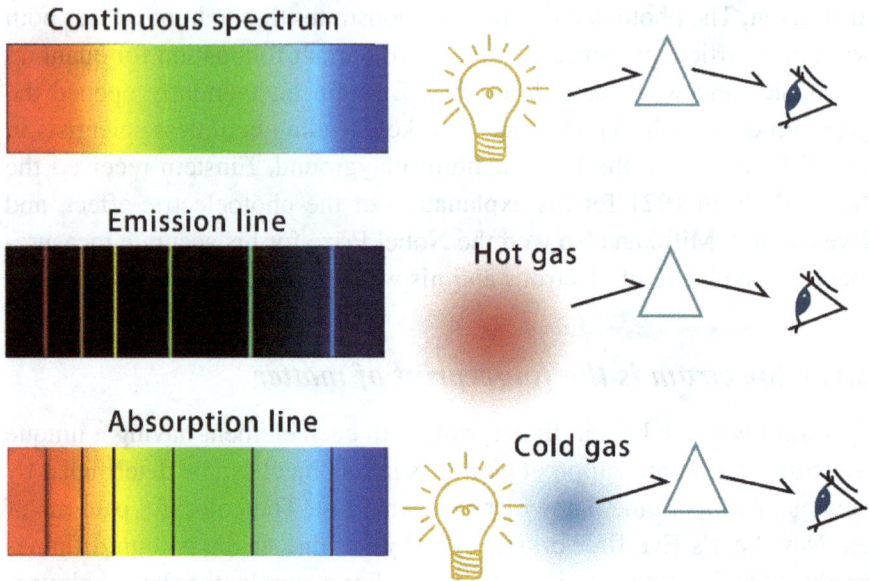

Figure 2.4.   When the light emitted from a heated filament is passed through a triangular prism, it produces a continuous spectrum. However, the spectrum obtained from the light emitted by a hot, rarefied gas consists of a series of discrete bright lines. In addition, when light passes through rarefied gas, it produces a spectrum consisting of discrete dark lines. The former is an emission spectrum, while the latter is an absorption spectrum. Bohr's atomic theory provides an explanation for the origin of these spectra.

$$\frac{1}{\lambda} = R\left(\frac{1}{n^2} - \frac{1}{n'^2}\right) \quad n = 1,2,3\cdots \quad n' = n+1, n+2, n+3\cdots$$

Later, Niels Bohr proposed the famous Bohr's atomic model to provide a theoretical explanation for this empirical formula.

2.2 The evolution of the atomic model

2.2.1 *Crack the door of the atom*

J.J. Thomson was the first to discover structure of the atom. In 1897, he discovered negatively charged electrons much smaller than atoms in experiments with cathode rays. Since atoms are electrically neutral but electrons are negatively charged, he thought that there must be other matter with the same amount of positive charge in an atom. Therefore, Thomson proposed the plum pudding model of the atom, which overturned Dalton's idea that the atoms that make up matter are indivisible solid spheres. The plum pudding model let scientists know that the atom is like a plum pudding, with raisins (electrons) inside, and there may be more. This led many scientists to devote themselves to the study of the internal structure of the atom. In 1909, Ernest Rutherford found some alpha particles bouncing off after striking gold foil. For about every 8,000 alpha particles, one particle was deflected by a large angle (even more than 90°), while the others passed almost straight through. This experimental result was very surprising, and Rutherford called it "the most incredible event that has ever happened to me in my life." How could an alpha particle, which is positively charged, be deflected so much by an atom? It is like shooting a pudding with a bullet and the bullet bouncing back. Such a large deflection of alpha particle cannot be understood according to Thomson's plum pudding model. The raisins (electrons) in the pudding also can not cause this phenomenon. The only possibility is that there is a non-uniform high-density structure inside the atoms. Rutherford suggested that the positively charged matter in an atom is concentrated in a small region, now known as the nucleus of the atom. The alpha particle bounced off when it hit the nucleus. And, the electrons moved outside the nucleus. The electrons in the atom were attracted by the positively charged nucleus due to the Coulomb force, thus orbiting the nucleus like a planet. This description is known as the Rutherford model of the atom. Rutherford's experiment laid the foundation for the modern atomic model.

According to Rutherford's model, the atomic structure is similar to the solar system, except for the gravitational force, which is replaced by the electromagnetic force. But, at that time, this planetary model of the atom had two serious problems that classical physics could not solve:

1. According to classical electromagnetic theory, any accelerating charged particle will emit electromagnetic radiation. Therefore, electrons in atoms will radiate electromagnetic waves of the same frequency as that of an electron's orbital motion, resulting in the

reduction of the electron's energy due to radiation energy loss. This energy loss will cause the electron's orbital radius to shrink rapidly, and electrons will collapse into the nucleus in a very short time. If the electrons in the atoms are really orbiting the nucleus, all the atoms in the universe are unstable and cannot exist.

2. Because the orbital radius of a planet can vary continuously, the model predicts that the radiation emitted by hydrogen atoms should have a continuous spectrum, which is inconsistent with the discontinuous spectrum observed experimentally. In the model of macroscopic planetary motion based on Newtonian mechanics, an artificial satellite orbiting the earth can be launched to any orbital altitude. In other words, the orbital radius of a satellite is continuous, so the energy of the satellite's motion is also continuous. If the earth is reduced to the size of an atom, the planetary model cannot explain why energy is discontinuous in the microscopic world.

Due the contradiction between old theories and new experimental facts, finding a reasonable explanation in line with the experimental facts and establishing a quantum atomic model became the Holy Grail for scientists at that time. Niels Bohr believed that the laws of macroscopic phenomena could not be used to infer the behaviour of atoms in the microscopic world and that new ideas were needed to understand atoms. Inspired by Balmer's empirical formula for the hydrogen spectrum in the visible light region, Bohr introduced the concept of light quanta, developed by Einstein, into the planetary model of the atom. Bohr described the atom as a small mechanical system, the main characteristics of which are similar to that of macroscopic planetary systems; but, he stated that the electron's orbit must meet a specific quantization condition. Bohr's planetary model of the atom solved the problems in one stroke. Bohr's planetary model of the atom not only solved the mystery of discontinuous spectral lines but also provided explanations for the stability of atoms and many periodic trends of elements in the periodic table. In 1913, Bohr published three papers with two main postulates:

1. **Stationary orbit**: The orbital radius and energy of electrons are quantized. Electrons do not radiate electromagnetic waves when they

travel in stationary orbits of certain discrete radii. In this way, electrons will not lose energy and crash into the nucleus. This postulate solves the instability problem of Rutherford's planetary atomic model.

2. **Energy level transition**: The allowed electron energies corresponding to stationary orbits are called energy levels. The lowest energy level is called the ground state and the high energy levels are called excited states. An energy level transition is the transition of an electron between two different energy levels. If an electron jumps from a high-energy orbit to a low-energy orbit, a photon of a specific wavelength will be emitted. Conversely, if an electron is to jump from a low-energy level to a high-energy level, it needs to absorb a photon of a specific wavelength.

Bohr's atomic model provides an explanation for the origin of discontinuous spectra and provides a derivation of Rydberg's empirical formula, which had a profound impact on atomic physics. The evolution of atomic model is summarized in Figure 2.5. Bohr's two assumptions of "stationary orbit" and "energy level transition" later became the basic concepts of quantum mechanics, and were the starting points for the abandonment of Newtonian causality. The Bohr planetary model of the atom captured the essence of the microscopic atomic world. In 1978, Paul Dirac, in his famous book *Directions in Physics*, said, "When I was a student in Bristol, I did not hear anything about this theory of Bohr; it was only when I went to Cambridge as a research student that I was told about it, and it then opened my eyes to a new world and a very surprising world." He also stated, "I still remember very well how strongly I was impressed by this Bohr theory. I believe that the introduction of these ideas by Bohr was the greatest step of all in the development of quantum mechanics."

Theoretical achievements, including the introduction of the quantum concept by Planck, Einstein's proposal of light quantum to explain the photoelectric effect, Bohr's atomic model, and subsequent refinements to the Bohr's model by Arnold Sommerfeld comprise what is called old quantum theory. The old quantum theory could indeed describe some simple phenomena by adding heuristically postulated quantization conditions on the basis of classical physics. However, it was still unable to

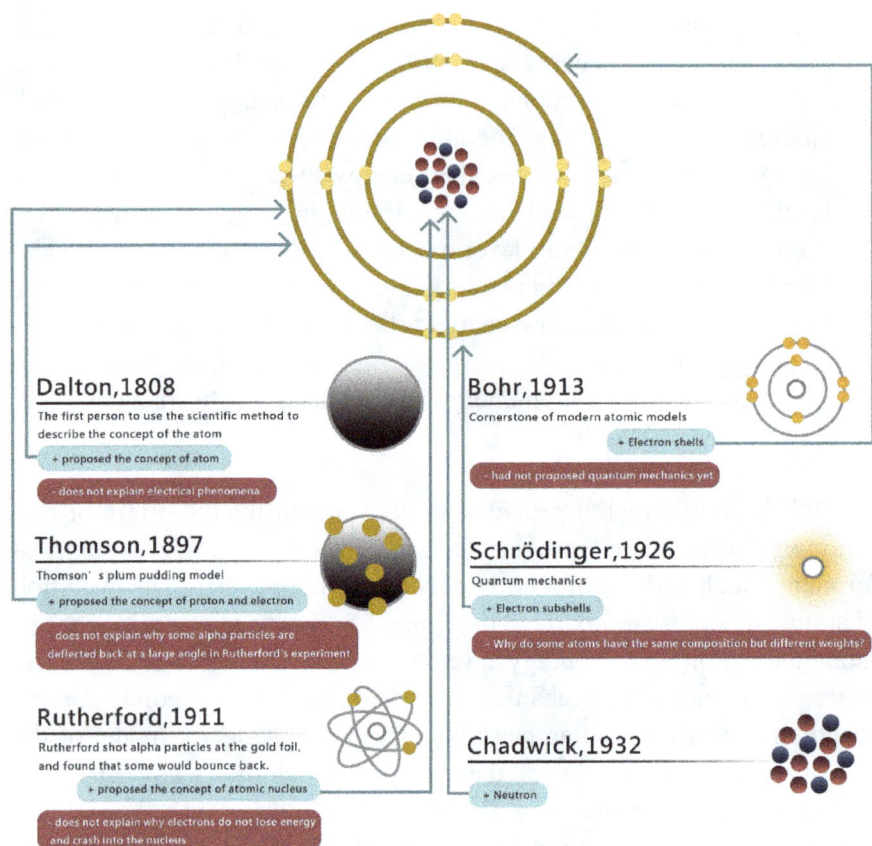

Dalton,1808

The first person to use the scientific method to describe the concept of the atom

+ proposed the concept of atom

- does not explain electrical phenomena

Thomson,1897

Thomson' s plum pudding model

+ proposed the concept of proton and electron

- does not explain why some alpha particles are deflected back at a large angle in Rutherford's experiment

Rutherford,1911

Rutherford shot alpha particles at the gold foil, and found that some would bounce back.

+ proposed the concept of atomic nucleus

- does not explain why electrons do not lose energy and crash into the nucleus

Bohr,1913

Cornerstone of modern atomic models

+ Electron shells

- had not proposed quantum mechanics yet

Schrödinger,1926

Quantum mechanics

+ Electron subshells

- Why do some atoms have the same composition but different weights?

Chadwick,1932

+ Neutron

Figure 2.5. Evolution of atomic model: Dalton proposed that atoms are the smallest particles that cannot be divided. Thomson suggested that atoms are like watermelons, with electrons, like watermelon seeds, scattered throughout the atom and embedded in the positively charged substance, which is known as the plum pudding model. Rutherford proposed the existence of a tiny, heavy, positively charged atomic nucleus within the atom. Bohr proposed that electrons behave like planets but only orbit the atomic nucleus in specific orbits. Schrödinger's wave equation led to the interpretation that electrons do not have definite orbits and their behaviour is probabilistic. Chadwick further suggested that atomic nuclei also have internal structures.

explain most of the more complex situations. For example, it failed to explain the anomalous Zeeman effect. It also had difficulty explaining multi-electron atoms. It was clear that the old quantum theory still needed further revisions.

2.3 The rise of quantum mechanics

2.3.1 *The advent of quantum mechanics*

In the 1920s, Germany was defeated in the First World War and was deeply discontented and also puzzled about reason why they will lose the war. Why did Germany lose the war even after working so hard? German society generally began to doubt the correctness of causality. Ironically, Germany's defeat was related to the Spanish flu pandemic. After German troops stormed enemy trenches in March 1918, they were infected with the virus from wounded soldiers who remained in the trenches. From March to August, due to influenza and war casualties, the German army lost at least 30% of its troops, about 800,000 troops. As a result, the German army was demoralized and many soldiers deserted the army. Kaiser Wilhelm II also fled abroad. Germany therefore sought peace to end the war. At that time, German political circles and society were filled with criticism of causality, which profoundly affected scientific thinking. In Newtonian mechanics, as long as the initial conditions and boundary conditions are given, scientists can determine the state of an object at any time through Newton's equations of motion. From the point of view of causality and classical mechanics, if you exert more force and work harder, you can reach the destination faster; but, the defeat of Germany showed that causality was flawed. At the time, physicists also happened to notice experiments which showed that classical physics has many flaws and deficiencies when applied to microscopic systems. The gradually developed quantum theory coincided with the political needs for non-causality of the German political authorities and society.

The development of quantum physics is mainly concerned with one question: Is matter a wave or a particle? There are two routes to the development of quantum mechanics. One route is wave mechanics. The concept of the matter wave was proposed by Louis de Broglie. Later, Schrödinger introduced the wave function and proposed the Schrödinger equation, which gave the matter wave a definite mathematical meaning. The other route is the matrix mechanics developed by Heisenberg and Bohr's Copenhagen school. These two different routes lead to the same goal, both of which can describe the motion of microscopic particles. Interestingly, these two routes are not only different in mathematical formalism but more importantly, the philosophies behind them are also contradictory. However, they have been proved to be completely mathematically equivalent.

2.3.2 *Wave mechanics*

Einstein's photon hypothesis involves directly assigning particle properties to light, which is known to have a wave nature. De Broglie proposed the matter wave hypothesis in his doctoral thesis in 1923, suggesting that wave–particle duality should not be a feature specific to light, but a general feature of any particle. The concept that particles exhibit wave-like properties is called matter waves. The contradiction between the particle nature and the wave nature of matter is similar to the controversy between geometric optics and wave optics in Newton's time, which is completely determined by the relative magnitude between light's wavelength and the size of the object. De Broglie boldly assumed that there should be a similarity between the principles of mechanics and optics, and tried to use this similarity to establish a new atomic-scale mechanical theory to describe the motion of microscopic particles. It is worth mentioning that although de Broglie is known for his matter wave hypothesis, his early interest was history and his first degree was a Bachelor of Arts (BA) degree in history.

The proposal of matter waves not only established the field of wave mechanics but also provided a clear physical picture for Bohr's stationary orbits. Considering the matter wave associated with an electron, the stationary orbits in Bohr's atomic model are orbits that satisfy the standing wave conditions. A standing wave literally refers to a wave that stands in space. An ideal standing wave does not travel but oscillates in place without losing energy. Plucking a guitar string creates standing waves with two fixed ends on the string. Changing the position of the finger changes the wavelength of the standing waves, and the guitar will emit sounds of different frequencies according to the different positions of the fingers (Figure 2.6 (a)). Since the string length must be an integer multiple of half a wavelength to form a standing wave, the wavelengths of different standing waves on a string are discrete, which is precisely in line with the quantum concept.

(a)

Fundamental mode

Second harmonics

Third harmonics

Fourth harmonics

(b)

Circumference is not an integer multiple of half-wavelength, unable to form a stable orbit of a standing wave.

Circumference is an integer multiple of half-wavelength, able to form a stable orbit of a standing wave.

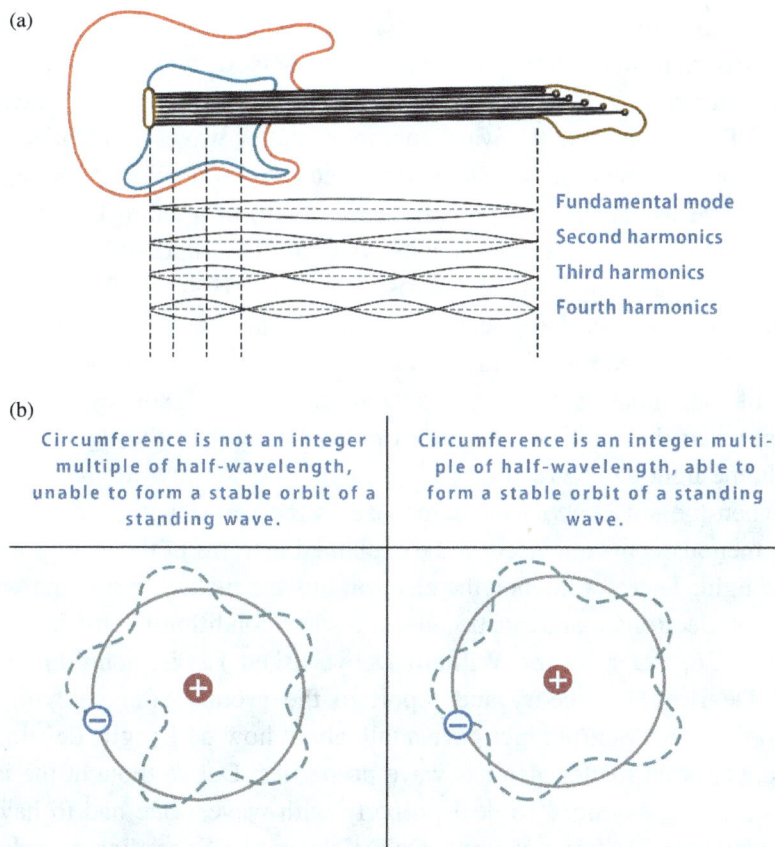

Figure 2.6. (a) The guitar will emit sounds of different frequencies according to the different positions of the fingers. To form a standing wave, the length of the string must be an integer multiple of half a wavelength, resulting in discrete wavelengths of different standing waves (also called different modes) on the string. (b) For a possible stationary orbit of electrons, electron matter waves must be able to form closed standing waves within the orbit. In the left-hand figure, since a closed standing wave cannot be formed within the orbit, it is not a stationary orbit of electrons.

In 1927, Clinton Davisson and Lester Germer shot electrons at a nickel crystal and observed patterns of electrons being scattered by nickel atoms. They unexpectedly found that the electron scattering pattern was the same as the patterns of the X-ray scattering experiments on crystals. The interference of electrons scattered by the nickel crystal, a characteristic phenomenon of waves, is important evidence for the wave–particle duality of electrons. Is it possible that tennis balls served by players in tennis matches are waves? If so, how have we not seen any wave phenomenon in tennis? This is because the wavelengths of matter waves associated with macroscopic objects are too short to be detected at all. For example, the matter wave wavelength (de Broglie wavelength) of a tennis ball moving at a speed of 100 km per hour is about 10^{-34} m. It is impossible to observe such a short wavelength in the macroscopic world. However, in the microscopic world, the de Broglie wavelength of an electron with a kinetic energy of fifty electron volts is about 1.7×10^{-10} m, which is on the same order of magnitude as the lattice spacing of solids. In single-slit experiments, when the width of the slit is much larger than the light's wavelength, the light can be regarded as a straight line passing through the slit; but, when the slit is about the same size as the wavelength, a diffraction phenomenon occurs and needs to be explained in terms of the wave properties of light. Therefore, when the electron hits the nickel crystal, the wave nature of electrons manifests because the wave condition is satisfied.

In 1926, Peter Joseph William Debye asked Erwin Schrödinger to study De Broglie's theory and report to the group. After studying de Broglie's thesis, Schrödinger gave a talk about how de Broglie developed the concept that matter also has wave properties. Debye thought the idea was too naive, saying, "To deal properly with waves, one had to have a wave equation." After receiving Debye's remark, Schrödinger took de Broglie's thesis along to study at leisure when he and his mistress went on a vacation to the Alps for two and a half weeks. Two and a half weeks later, the world changed dramatically! Schrödinger presented a nonrelativistic wave equation in the next group talk soon after he returned from the Alps, which is the famous "Schrödinger wave equation." If we think of electrons as waves, just like clouds in the sky (this is also the origin of the term "electron cloud"), the Schrödinger wave equation is the equation of motion for the wave function of electrons. The electron's wave function in the Schrödinger equation is represented by the Greek letter Ψ, which describes the hydrogen atom concisely and elegantly. Just like the importance of Newton's laws of motion in classical mechanics, everyone who studies quantum mechanics must understand the well-known "Schrödinger wave equation." Years later, Bloch asked Debye if he regretted suggesting that Schrödinger work on the wave equation instead of deriving it himself. Debye responded generously, "Well, wasn't I right?"

The world of Newtonian mechanics is a world of real coordinate space where the result of any physical measurement is a real number. However, in quantum mechanics, it is necessary to use a wave function containing both real and imaginary parts to describe the quantum phenomena because of the wave nature of microscopic particles. The wave function in the "Schrödinger wave equation" is a complex one rather than a real one. What is the physical meaning of it? German physicist Max Born understood the wave function as a probability amplitude, and that it is the square of its absolute value (Modulus square of complex number) that really corresponds to the measured probability distribution of particles. Born believed that the matter wave itself is actually a probability wave. Simply speaking, the behaviour of particles in spacetime is probabilistic. Particles can be in multiple positions at the same time, which is the concept of the superposition state. This probability wave interpretation was completely unacceptable for scientists who generally believed in determinism and locality at that time. Even Schrödinger himself refused to accept the interpretation that the wave function in the equation he discovered was the probability of particles. Einstein even wrote the famous statement "God does not play dice" in a letter to Born.

2.3.3 *Matrix mechanics*

The Copenhagen school was a major school of interpretation of quantum mechanics founded at the University of Copenhagen, Denmark, with Bohr's complementarity principle, Born's probability wave, and Heisenberg's uncertainty principle as the three main pillars. The Copenhagen school believed that the goal of physical research was to understand observable phenomena, rather than make conclusions based on unobservable or pure reasoning.

Werner Karl Heisenberg's style of doing physics was different from that of most theorists. He believed that only when the observable quantity in experiments was introduced into the physical theory could it be truly meaningful, and all theories must come from experimental results. Heisenberg argued that since the details of the electron motion in atoms cannot be directly observed, it is pointless to spend time discussing the orbits of electrons in atoms. Since only the specific frequency corresponding to the energy radiated when an electron makes a transition can be observed in the experiment, the experimental observable quantity is the energy level difference. Therefore, Heisenberg studied the behaviour of electrons based on the energy level difference. In May 1925, he attempted to describe atomic systems using only relations between observable quantities (such as spectral line position and intensity), rather than the concept of electron orbits in Bohr's theory. On July 7, Heisenberg was recuperating from hay fever on a small North Sea island. On this pollen-free island, he happily read Goethe's poems to inspire thinking about atomic spectral lines. It was almost three o'clock in the morning when he finally finished his calculations. He was deeply shocked. He was so excited that his drowsiness completely vanished. He thus left the room and sat on a rock to wait for the sunrise. Heisenberg presented the results he concluded from spectroscopic experimental data to his teacher Born, who found that the resulting mathematical form could be described by matrix operations. Matrix mechanics, which describes the behaviour of microscopic particles, was born.

2.3.4 *The uncertainty principle*

Heisenberg's matrix mechanics cannot describe the trajectories of electrons in atoms, because matrix mechanics is a theory based on a mathematical form deduced from observable results in experiments, and it makes no attempt to understand the position or momentum of electrons in atoms. Einstein immediately questioned the theory: "You do not mention the path of the electron at all, Heisenberg. But yet when you look in a cloud chamber the electron's track can be observed quite directly. Don't you think that it's strange to say that there is a path for the electron in the cloud chamber, but there is no path for the electron in the atom?" After much thought, Heisenberg finally figured out that the trajectory observed in the cloud chamber was caused by water droplets condensed after water vapour was struck by electrons. Since the volume of each drop was much larger than the actual volume of electrons, the electron trajectory observed in the cloud chamber was very different from the precise trajectory of electrons. After careful calculation, Heisenberg proposed the famous uncertainty principle to respond strongly to Einstein: The product of uncertainty in position and uncertainty in momentum is greater than or equal to Planck's constant. To put it in everyday terms, the uncertainty principle states that the position and momentum of a particle cannot be simultaneously determined with arbitrarily high precision. This response brought fame to the Copenhagen school. It is difficult to understand the uncertainty principle in quantum mechanics from the perspective of classical physics. In classical physics, the objects being analyzed possess independent and deterministic properties.

Heisenberg thought that the uncertainty relation between position and momentum originated from the disturbance the observer brought to the measured system. When measuring a physical quantity, the act of measurement will affect the measured result because the observer interacts with the measured system in the measurement process. Heisenberg explained it with a thought experiment in which an observer tries to measure the position of an electron using light rays. If one wants to determine the position of a particle more accurately, one must use light with a shorter wavelength to measure it. But, the shorter the wavelength, the stronger the energy of the photon, and the more it will change the particle's momentum when colliding with the particle, so there will always be a trade-off between the measurement precision of the position and momentum. The uncertainty principle implies that the electron motion in the atom can only be known within a certain acceptable range, rather than perfectly moving in its orbit around the nucleus.

Bohr once criticized Heisenberg's thought experiment for not being revolutionary enough. He believed that electrons in atoms should be described by probability distributions, whether measured or not, and that electrons simply do not have what Einstein called the "path." Indeed, it has been experimentally confirmed that the uncertainty relation discovered by Heisenberg does not come from the disturbance introduced by the observer. Instead, the uncertainty principle is inherent in the wave description of particles and is closely related to wave–particle duality, as Bohr suggested.

2.3.5 *There are two sides to everything — The complementarity principle*

Whether light behaves as a wave or as a particle has been debated for hundreds of years since the time of Newton. In the first wave–particle debate, Newton's particle theory was more generally accepted than Huygens's wave theory. However, during the second wave–particle debate, Augustin-Jean Fresnel successfully explained the diffraction phenomenon of a bright spot appearing in the shadow of an opaque disk when a collimated beam of light shone on it, which strongly supported the wave theory of light. The third wave–particle debate started after Einstein's photon hypothesis. After so many experiments and discussions of principles, who would win the war between wave theory and particle theory? Let us recap: In classical physics, electrons are particles, and light is a wave, which are completely different. Einstein proposed that light has particle properties, and de Broglie proposed that electrons have wave properties. Heisenberg treated the behaviour of electrons by matrix operations, and Schrödinger formulated the equation of motion for electrons considering the wave function of electrons. The dispute between the opposing views of "particle" and "wave" seems to have been repeated and inconclusive. Bohr's complementarity principle gave a statement about the situation: The wave description and the particle description are complementary rather than mutually exclusive. The real situation of the universe is that "there are two sides to everything." Bohr believed that whether an object showed wave character or particle character depended on how it was measured. As shown in Figure 2.7, depending on the viewing angle, the projection of a cylinder on the screen can be a circle or a square. So, what exactly is an electron? We should first ask what kind of measurements we perform on the electron. The key lies in the nature of the measurement. Once the measurement method is determined, the electron will appear as a wave or particle according to the measurement setup. Like the parable of the blind men and the elephant, each measurement can only reveal part of the electron. Only when all aspects are taken into account can a complete description of things be obtained, which is the concept of "complementarity" suggested by Bohr. The wave feature and particle feature are like two

Figure 2.7. A cylinder's projection on a screen varies significantly depending on the viewing angle. A complete and accurate description of the object can only be achieved through repeated observations that produce consistent outcomes.

sides of the same coin; both sides are part of the coin, and both are indispensable. Bohr's complementarity principle perfectly settled the long-standing wave–particle debate.

In 1927, Bohr conceived the complementarity principle while on a ski holiday in Norway. In 1937, Bohr was invited by Chinese physicist Zhou Peiyuan to give lectures in China. When enjoying the Peking Opera in Beijing, he saw Jiang Ziya holding the Taiji flag in his hand. Bohr admired the Taiji diagram, and he believed that the "complementarity principle" and wave–particle duality were all implicit in the yin and yang of Taiji. In 1947, the King of Denmark awarded Bohr "the Order of the Elephant," the highest honour that only members of the Danish royal family and heads of state could obtain, in recognition of his outstanding scientific contributions. Bohr designed his own coat of arms with the Taiji diagram (Figure 2.8), which read "Contraria sunt complementa" ("opposites are complementary" in Latin). Dirac described Bohr as "the deepest thinker I had ever met" and praised him as "The Newton of atomic theory."

Figure 2.8.   Bohr designed his coat of arms with the yin–yang Taiji diagram, expressing that opposites are complementary.

2.4 Concluding remarks

Is the correct description of the quantum world Schrödinger's wave mechanics or Heisenberg's matrix mechanics? Is matter a wave or a particle? Although the starting point and underlying philosophical thinking of wave mechanics and matrix mechanics are entirely different, Schrödinger and Pauli (Wolfgang Ernst Pauli) proved that the two methods are completely mathematically equivalent. Furthermore, British physicist Paul Dirac combined matrix and wave mechanics perfectly when studying relativistic quantum mechanics and proposed the famous Dirac equation. The novel principles of quantum mechanics have changed the way people think about the natural world.

The Copenhagen school followed empiricism and believed that understanding the underlying mechanisms was not necessary. It was opposed by physicists headed by Einstein who believed in causality. Einstein strongly criticized the Copenhagen school's interpretation of the wave function, the uncertainty principle, and the complementarity principle. Einstein could not accept the physical world without strict causal laws. He believed that all events cannot happen randomly and there must exist an objective reality. However, Bohr believed that meaningful reality is based on the measurement method, so the existence of the moon is related to observation. After many debates, Bohr always successfully fended off challenges from Einstein.

In May 1935, Einstein, Boris Podolsky, and Nathan Rosen jointly challenged the completeness of quantum mechanics again and proposed the famous thought experiment of the EPR paradox. The EPR paradox does not question the correctness of quantum mechanics, but uses two entangled particles to argue that quantum mechanics is not complete. The EPR paradox is based on the two common ideas at the time, locality and realism. Simply put, the principle of locality states that events in a certain area cannot be transmitted to other areas at a speed exceeding the speed of light. The principle of locality does not allow the long-distance correlation exhibited by two entangled particles, which Einstein called "spooky action at a distance." On the other hand, realism holds that what is observed in experiments is independent of how it is observed and the act of observation. In other words, the moon's existence has nothing to do with whether people watch it. The view that nature should conform to the principle of locality and realism is known as local realism. Einstein believed that any complete physical theory must fulfil local realism.

Six weeks later, Bohr responded to the EPR challenge with an article in the October issue of *Physical Review*. Bohr countered that the statement "without in any way disturbing a system" in the EPR article's assumption about physical reality was fallacious, and therefore the inference of the EPR thought experiment that "quantum mechanics is not complete" was not valid. Following the EPR paper, Schrödinger also proposed another famous thought experiment "Schrödinger's cat" to support Einstein, trying to prove the incompleteness of quantum mechanics and highlight the absurd quantum measurement. In quantum mechanics, once two particles get close and become entangled with each other, they will lose their original individuality and become a whole state of the two particles. In the future, no matter how far the two particles are separated in space, as long as the entangled state is preserved, the wholeness will not disappear. This phenomenon is called quantum entanglement by Schrödinger.

Quantum entanglement is a purely quantum effect, which does not exist in the classical world, and is thus most difficult to accept and understand. Einstein and Bohr were not persuaded by each other throughout their lifetimes. After Einstein passed away, the anti-Copenhagen school lacked a leading voice, and the debate between classical mechanics and quantum mechanics dissipated, with the EPR paradox remaining unresolved. Is quantum mechanics still incomplete, or is there really "spooky action at a distance" in the quantum world? In 1964, Irish physicist John S. Bell proposed Bell's inequality, providing an experimental method to test quantum entanglement. However, it was not until the 1980s that actual experiments were performed. In 2015, the experiment conducted by Dutch scientists was the first to rigorously refute "local realism" and confirm that "Spooky action at a distance" does exist. The journal *Nature* then reported this result under the title "Death by experiment for local realism," which states, "A fundamental scientific assumption called local realism conflicts with certain predictions of quantum mechanics. Those predictions have now been verified, with none of the loopholes that have compromised earlier tests." Ironically, the challenge of the EPR paradox turns out to be a vindication of quantum mechanics. The important characteristics of the microscopic quantum world — uncertainty, superposition, nonlocality, and entanglement — have since become unshakable truths.

After two dark clouds poured rain on the arid classical world, many creative ideas sprung from the classical soil. As a result, quantum technology began to flourish. Although humankind's understanding of quantum technology still needs to develop, the fifth Solvay Conference has been positioned as the most knowledgeable and historic conference. From the perspective of the development process of quantum mechanics, the Copenhagen school, which emphasized experimentalism, correctly constructed the basis of current quantum technology based on observable results. However, in exploring the deep principles of the microscopic world, it was de Broglie, Schrödinger, and Einstein who continued to raise challenging questions that made the quantum theory more complete. The focus of the debate between Bohr and Einstein at the beginning was the inherent self-consistency and uncertainty problem of quantum mechanics. Bohr was on the winning side in the two big debates of the fifth and sixth Solvay conferences. Einstein's objection of "God does not play dice" underlines the probabilistic characteristic of quantum theory. His challenge to the uncertainty principle — the "photon box" thought experiment — also highlights this important property of quantum systems. Due to political tension, Einstein remained in the United States and did not attend the seventh Solvay Conference. However, starting from quantum mechanics, he still proposed the famous EPR paradox in an attempt to prove that quantum mechanics was incomplete. What is interesting is that Einstein's last challenge, "spooky action at a distance," actually led to the confirmation of quantum entanglement! These profound criticisms, which had lasted for more than 10 years, furnished the Copenhagen school with endless motivation to think and energy to research. The debate on quantum mechanics between Einstein and Bohr has been called the 20th-century battle of ideas. Their brainstorming with each other was like another manifestation of complementarity, and thus completed the most important quantum knowledge in human history.

Quantum mechanics not only explains atoms but can also describe how atoms are bonded into molecules, the formation of crystal structures, and the electronic states and energy bands in solid materials. Studying crystalline materials through quantum mechanics can unravel the secrets behind many material phenomena, including the principle of semiconductors. The results of applying quantum concepts to life are far beyond imagination, especially the achievements related to condensed matter physics. Without quantum mechanics, there would be no high-tech industry today. Without understanding black-body radiation, there would be no thermal imaging cameras in today's airports. Similarly, there would be no solar cells today without the understanding of the photoelectric effect. As stated in this chapter, the ideas and principles of quantum science were indeed developed in Europe. One of the main issues discussed at the 2018 EU Quantum Flagship Kick-Off Conference was why the "First Quantum Revolution" did not appear in Europe and why the United States mastered the main application achievements of quantum mechanics for a long time. The primary objective of the EU Quantum Flagship Kick-Off Conference was to enable the "Second Quantum Revolution" to be initiated and completed by Europe.

This question is similar to a historical question in the modern East. Why did modern science not arise in China? Where do breakthrough scientific thinking and technological innovation come from? Many similar ideas of modern science appeared in China very early, but they never became a systematic modern science, let alone a technological revolution. For example, geometric optics and pinhole imaging have been recorded as early as the time of Mozi, and the use of bronze mirrors to focus sunlight and light a fire existed as early as the Zhou Dynasty. But, why did lenses, microscopes, and telescopes not appear in China, not to mention modern biology with microscopes and modern astronomy with telescopes? When Bohr saw the Taiji diagram, he felt that the concept of yin and yang had something in common with his "complementarity principle" and wave–particle duality. Wang Yangming, a philosopher of the Ming Dynasty, once said, "When you haven't looked at this flower, this flower and your heart go to silence together; when you look at this flower, the colour of this flower becomes clear for a moment." This kind of idealist thinking is also very similar to how quantum measurement works, but why did quantum science not appear in China? Perhaps Heisenberg's uncertainty principle is the answer: There will always be a trade-off between the development of thought and technology. Looking back at the development of quantum mechanics, we can understand that whether starting from the experimental evidence or essence, as long as the attitude of pursuing knowledge is modest enough, coupled with the same purpose of exploring truth and perseverance, we will always reach the same goal of truth.

Chapter 3

High-Dimensional Hilbert Space

Only extreme quietness can detect the movement of the crowd, and only empty space can absorb all realms.

— Su Shi (1037–1101)

We must know, we must know.

— David Hilbert (1862–1943)

3.1 Quantum state and complex hilbert space

3.1.1 *The wave nature of quantum mechanics*

The previous chapter introduced a lot of history and experiments related to the early development of quantum theory, including some quantum phenomena that classical physics cannot explain. For this reason, scientists have proposed many explanations and properties that defy intuitive understanding, such as probability waves. To understand probability waves, we must first review the concept of waves in classical physics.

Waves can be seen everywhere in daily life, whether it is the water waves generated by the ripples on the lake surface, the sound waves generated by the hawkers shouting loudly on the road, or the sunlight that catches your eye when you get up in the morning. People have studied waves in the natural world for a long time. Scientists have attempted to describe these phenomena using systematic mathematical language,

55

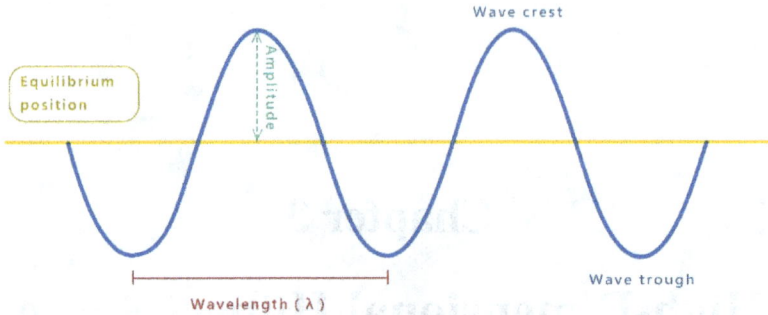

Figure 3.1.   Wavelength, amplitude, wave crest, and wave trough.

leading to the development of different branches of physics, such as geometric optics and electromagnetism. The mobile phone that everyone uses
every day transmits information through radio waves, and microwaves are
used to cook food in a microwave oven at home. Both are applications of
waves in technology. Intuitively, a wave can be understood as a geometric
mechanical object with amplitude, phase, and frequency. Each point on
the wave vibrates periodically over time and transfers energy in the way
that one point drives another point. Figure 3.1 shows a periodic wave at a
given time, where the shape of the wave repeats after a certain distance,
known as the wavelength (λ). The maximum value of the wave is called
the amplitude. The highest position of the wave is called the wave crest,
and the lowest point is called the wave trough. The speed at which a wave
travels in space is called the wave velocity (v), and the period T is the time
it takes for a complete oscillation. Frequency f is the number of wave
cycles per unit of time, which is the reciprocal of the period T.

A special concept in wave phenomena is phase, which is simply the
relative difference between the starting positions of two waves. When the
two waves are identical and their peaks and troughs coincide, they are said
to have the same phase or zero phase difference. When the peaks of one
wave overlap with the troughs of another wave, it is called a phase difference of 180°. In other situations, the phase difference still can be determined according to the specific relation of two waves. Interestingly, when
several waves of the same frequency meet, interference will occur if the
relative phases of these waves are different. As shown in Figure 3.2, on the
one hand, when the relative phase of the two waves is the same or zero,

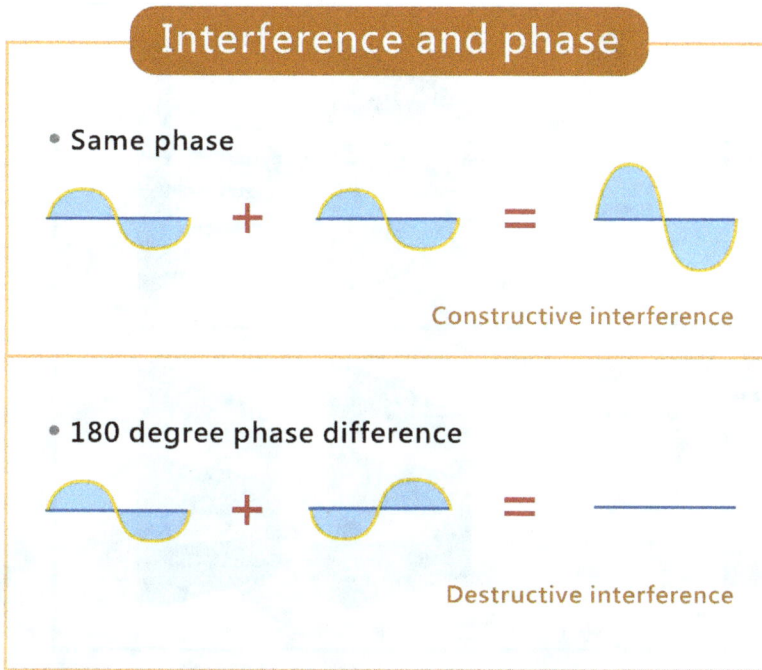

Figure 3.2.   Constructive interference and destructive interference of waves.

constructive interference occurs. In this case, the amplitude of the composite wave will be the addition of the amplitude of two original waves, thus becoming larger. On the other hand, if the relative phase of the two waves is 180°, destructive interference will occur instead. In this case, the two waves will cancel each other, leading to a reduction or complete disappearance of the amplitude of the resultant wave. Understanding waves' interference properties helps us grasp why wave properties are necessary in quantum mechanics. In the following, we illustrate the wave properties of quantum mechanics through the electron double-slit experiment.

The experimental setup of the electron double-slit experiment consists of an electron beam emitter, a piece of metal with two slits, and a screen with a detector that records each electron's position after passing through the slits. We intuitively expect that the distribution of landing points on the screen after the particle beam passes through the slits will be the combined result of the distributions produced when the two slits are opened

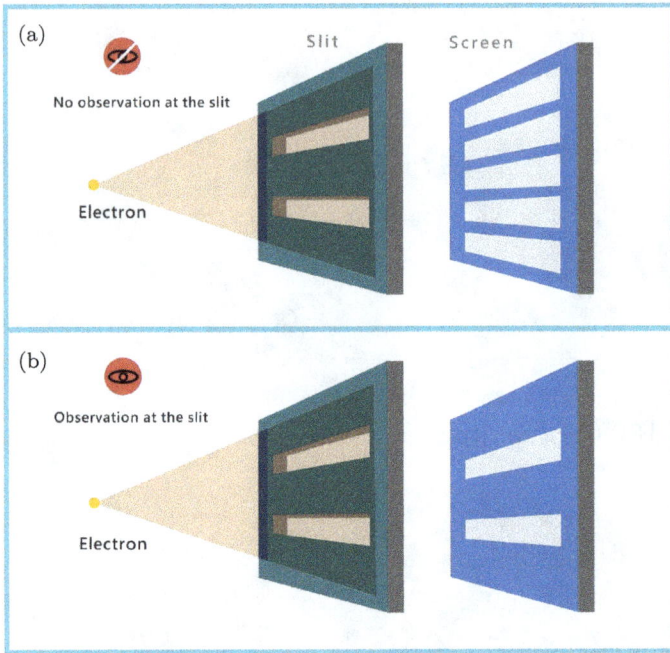

Figure 3.3. Schematic diagram of the electron double-slit experiment and the patterns observed on the screen. (a) When no detector is placed at the slits, and detection occurs only at the screen, the double-slit interference pattern of electrons is observed. (b) When a detector is placed at the slits, the interference phenomenon disappears. Only the combined result of the patterns produced when the two slits are opened individually is observed on the screen.

respectively, as shown in Figure 3.3(b). If the electron beam is replaced by a light beam, due to the wave nature of light, it will produce interference fringes with alternating bright and dark regions of different intensities when passing through the double slits. Contrary to our expectations, inter-ference fringes with alternating brightness also appear on the screen in the electron beam experiments. What's even more amazing is that when the electron beam intensity was further reduced to where only one electron passed through the double slit at a time, the double-slit interference pat-tern still appeared on the screen. If indeed only one electron passes through the double slit at a time, the electron should only be able to pass through one of the two slits in the macroscopic world. How is it possible, then, to observe an interference pattern? Could it be that the electron is somehow cloning itself and interfering with itself? The only reasonable explanation for the results of electron double-slit experiments is that electrons also have wave properties in the microscopic world. More precisely, electrons appear in space in the form of probability waves, as described by the wave function in the Schrödinger equation.

If electrons and objects exhibit wave characteristics, some readers may wonder why we don't observe wave phenomena in daily life. This is because objects in our daily life have a very large mass compared to electrons, causing the wave phenomena of probability waves to be too small to be observed with the naked eye. Therefore, classical mechanics is sufficient to describe the motion of objects in the macroscopic world.

The results of electron double-slit experiments show that objects are not only particle-like but also wave-like. In the world of Newtonian mechanics, all measurable physical quantities are real numbers, and they can all be measured precisely and independently. The measuring tools used to measure length, mass, and time in daily life can only yield real-number results. In other words, the world of classical mechanics described by Newton is a world of real numbers in a three-dimensional Euclidean space. In quantum mechanics, to describe the properties of probability waves of particles, it is necessary to use wave functions that contain both real and imaginary parts. Therefore, complex numbers must be introduced to describe the wave properties of quantum states. This is achieved by utilizing infinite-dimensional Hilbert Space in mathematics to describe the state space where quantum systems reside. The probability of a particle appearing in space is the square of the absolute value of its complex probability wave function.

3.2 Vector spaces and hilbert spaces

3.2.1 *Hilbert space*

David Hilbert is one of the most influential mathematicians in history. Hilbert was born in Konigsberg in 1862. He made significant contributions to various fields, including axiomatic geometry and number theory. In 1900, Hilbert put forward the famous 23 difficult mathematical problems

at the International Congress of Mathematicians, which inspired genera-
tions of students to devote themselves to the mathematics field. One of the
concepts proposed by Hilbert, the Hilbert space, has a significant impact
on axiomatic mathematics and the mathematical description of quantum
mechanics. Hilbert is often referred to as "the last all-rounder in mathematics." There are so many mathematical terms named after Hilbert, some
of which are even unknown to Hilbert. Hilbert once asked a colleague:
"What is Hilbert space?" His famous quote from his 1930 retirement
speech is still alive today: "We must know, we must know" (Wir müssen
wissen, wir werden wissen).

In mathematics, Hilbert space is a generalization of Euclidean space
that extends beyond the limited dimensions of two or three. All the concepts of plane geometry and three-dimensional space we learned in school
can be extended to Hilbert space. Hilbert space can be sort of regarded as
"infinite-dimensional Euclidean space." If Euclidean space is like a
banana, then Hilbert space encompasses all kinds of fruits. Although the
range of fruits is infinitely broad, the method of dealing with all fruits is
fundamentally similar. Hilbert space, like Euclidean space, is an inner
product space, with the concept of distance and angle, and the orthogo-
nality of vectors. From a mathematical point of view, Hilbert space has the
following characteristics: completeness, inner product property, and being
a vector space. In the following, we briefly introduce these three concepts:
completeness, vector space, and inner product.

(1) **Completeness:** There are many definitions of completeness in math-
    ematics. Here, we explain it in an intuitive way using the complete-
    ness of the number system.

    We all know that each point on the number line represents a real
    number, and no matter how close the two real number points are,
    there must be another real number between them. The number line
    can be densely populated by an infinite number of real number
    points, which is the completeness of the real numbers. However, this
    is not the case for rational numbers. Due to the existence of irrational
    numbers, all rational number points cannot fill the entire number
    line, which means that rational numbers are incomplete. For an
    incomplete space, its properties are like a piece of cotton fabric.

No matter how tightly woven, there will always be small water-absorbing holes. In contrast, a complete space is like a smooth plastic sheet with no water-permeable holes. Hilbert space has such a complete property.

(2) **Vector space:** Vector space is a concept derived from space in physics and geometry. In vector space, we can follow the notion of vectors and vector operations in middle school geometry. There are many daily-life examples of vector space. For example, John and Mary made an appointment to go to West Lake for dinner, but Mary got lost. In order to help Mary reach her destination, John must first know Mary's current location. John used WeChat (instant messaging software) to ask, "Where are you?" Mary replied, "I am at the southeast gate of Lingyin Temple in Hangzhou." There is only one Lingyin Temple in Hangzhou, so John can be very sure about the current location of Mary, for which the GPS coordinates are (30.237, 120.105). In the mathematical sense, knowing the GPS coordinate is equivalent to knowing the position vector of Mary relative to the origin, and the set of all position vectors is the vector space.

(3) **Inner product:** An operation between two vectors, called inner product, is introduced to specify the relative relationship between any two vectors in a vector space. The inner product allows the definition of the length of a vector and the angle between two vectors. In the above example, after knowing the location of Mary, John needs to determine the distance and relative direction between him and Mary in order to

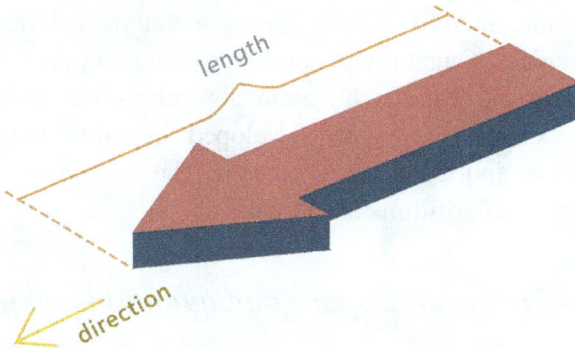

Figure 3.4. A vector consists of both length and direction.

meet her as quickly as possible. By calculating the inner product of his position vector and Mary's position vector in the vector space, John obtains the route and sends a message saying, "You have to walk 100 meters to the northeast, and then turn right and you can see me." Mary replied: "Okay, I'll come over right away!"

The application of concepts of vectors can be seen everywhere in everyday life. Here is another example: In today's convenient transportation, humans can often travel quickly between countries in just a few hours or days. In the sky, there are often hundreds or even thousands of aircraft flying at the same time. To accurately track these aircraft, it is important to use a coordinate system to describe their positions in the air.

By defining the position of the origin and the coordinate system, we can use vectors to represent the position of any airplane in space. In addition, the direction of velocity vectors allows us to determine the heading of the aircraft. In this way, air traffic controllers can constantly monitor the movements of the aircraft and provide appropriate support and guidance.

Through mathematics, human beings can analyze and deal with many complex problems encountered in daily life, thus greatly promoting the development of science and technology. In addition to everyday problems, we can also axiomatize the concepts of length and angle more specifically, extending them to physical systems and abstract complex spaces. This has led to the development of concepts in Hilbert space and quantum mechanics. Physicists employ Hilbert space to describe the spaces in which quantum states reside, thereby establishing a mathematical framework for quantum mechanics. Quantum computing is built upon the foundations of quantum mechanics. Various quantum algorithms based on principles of quantum mechanics have been developed, including the well-known Shor's Algorithm and Grover's Algorithm. The following chapters will introduce quantum algorithms in more detail.

3.2.2 *High-dimensional space and quantum computers*

The usual concept of space refers to a three-dimensional space composed of length, width, and height. Time itself has certain characteristics of dimensions. For example, the time axis can connect many three-dimensional spaces. Therefore, many people think that we live in 3+1-dimensional spacetime (also known as four-dimensional spacetime). However, time is very different from length, width, and height. For example, the unit of time differs from the unit of length. Additionally, time travels in a single direction and cannot be turned back. Therefore, it is not truly a four-dimensional space, but rather a composite image of many snapshots of three-dimensional space. Euclidean space can describe the three-dimensional space of our daily life very well. For ordinary people, "space" refers to the three-dimensional space in which they carry out their everyday activities. But for mathematicians, "space" belongs to the field of geometry in mathematics. In geometry, the difference between points, lines, planes, and solids is merely the difference in the number of dimensions defined. A world where movement occurs only along a straight line, forward and backward, is known as a one-dimensional world. In a one-dimensional world, everything is only allowed to move in one dimension, like pearls on a necklace. In a two-dimensional world, objects can move "forward/backward" and "left/right" within a plane, similar to pieces on a chessboard. In a three-dimensional world, objects can move "up/down," "forward/backward," and "left/right." The space in which humans live is a three-dimensional space. By extension, in an N-dimensional world, objects should be able to move in N directions. This kind of space is known as high-dimensional Hilbert space.

Hilbert space is needed when the spatial dimension of objects or properties of objects is no longer limited to two or three dimensions. The microscopic quantum world requires a high-dimensional space for a complete description. N 2-level quantum system lives in the 2^N-dimensional Hilbert space. This high-dimensional space is interwoven by many low-dimensional "sub-worlds," just like the three-dimensional world can be formed by stacking many two-dimensional planes. Each "sub-world" can only perceive the traces of the high-dimensional space vector projected onto it, so the dynamic movement in different "sub-world" is very different. Physicists have utilized the mathematical concept of Hilbert space to successfully describe the behavior of quantum phenomena in the microscopic world, just as three-dimensional Euclidean space provides a clear understanding of Newtonian mechanics.

3.3. Concluding remarks

When using a classical computer to simulate quantum phenomena, as long as the number N increases, the dimension of the Hilbert space will exponentially grow to a huge 2^N dimension. Consequently, simulating the vast Hilbert space of the Schrödinger equation on a classical computer would require an impractical amount of time. Richard Phillips Feynman proposed that if a quantum computer made of quantum bits (qubits) was used to simulate quantum phenomena, the computation time could be significantly reduced. A quantum computer operates within a high-dimensional Hilbert space generated by entangled qubits. High-dimensional space has boundless capabilities. In fact, ordinary people are always worried about living in a limited 3+1 space and are often confused by illusory projections from high-dimensional Hilbert space. Hilbert has long cleared all doubts (we must know, we will know) as long as we think the way of the high-dimensional sub-world, all troubles in the reality will automatically disappear. All operations on a quantum computer are performed in the Hilbert space rather than the digital world of 0s and 1s in classical computers. The ability to directly perform measurement in high-dimensional Hilbert space without the need to do any digital calculations is the major reason why quantum computers can outperform classical computers.

Chapter 4

Reversible Quantum Operations

Every year the flowers look the same as time goes by, but every year people change and get old.

— Liu Xiyi (651–679)

Time is life. It is irreversible and irreplaceable. To waste your time is to waste your life, but to master your time is to master your life and make the most of it.

— Alan Lakein (1938–)

4.1 Reversible and irreversible processes

One of the most attractive selling points of the 2020 blockbuster sci-fi movie *Tenet* is the exploration of the grandfather paradox through reciprocating time travel within multiple parallel universes. Many people cannot understand the complex parallel time–space plot in the movie, which sparked heated discussions on the Internet. Time-looping narratives have been explored in other sci-fi films like *Source Code* and *Groundhog Day*. However, Christopher Nolan was the first director to have the courage to handle the vast and intricate scene of simultaneous time travel in multiple spaces. The idea of time-looping within multiple spaces likely draws inspiration from reversible computing in the field of quantum computing. In the high-dimensional Hilbert space spanned by multiple qubits, quantum operations can operate in a reversible manner.

Unfortunately, life cannot be repeated, and we cannot reshape our destiny! When we take the high-speed train from Taipei to Kaohsiung to see friends and then back to Taipei, this spatial movement is a reversible process. But, things in the world are irreversible as long as they are related to time. We can never be young again. Many deep regrets in life cannot be reset like a video game. The only thing we can do is face our choices bravely and embrace life happily. This irreversible process exists everywhere in life. When ink drops into water, the whole glass of water gradually turns black. Turning the contaminated water back into clear water has never been seen except in magic tricks. The spread of the COVID-19 virus is also irreversible. Once the virus breaks through the lockdown and infects a community, achieving zero cases again becomes incredibly challenging. Things can only be prevented carefully beforehand.

To understand why the processes of the macroscopic world are always irreversible and why life cannot be reset and restarted, one must learn the basics of thermodynamics. Most results of thermodynamics are built upon the accumulation of empirical laws derived from experiments. The laws of thermodynamics define various macroscopic physical quantities such as temperature, internal energy, entropy, and pressure. Thermodynamics is the subject that describes the relationship between these physical quantities. Physical quantities in thermodynamics are the result of the average behaviour of many classical particles. The study of thermodynamics necessitates the use of probability distributions, but the physical interpretation of probabilities is entirely different from that of quantum mechanics. Many engineering applications are closely related to thermodynamics. Heat is a form of energy transfer. The atoms or molecules of an object transfer energy from a higher-temperature region to a lower-temperature region through random motion and collision. Thermodynamics mainly studies the equilibrium states of matter and the processes involving near-equilibrium states, especially focusing on energy exchange between a system and the external environment.

Physicist Lídia del Rio once described thermodynamics as follows: "If physical theories were people, thermodynamics would be the village witch." Unlike other physical theories, which aim to understand the principles and mechanisms of phenomena in the universe, thermodynamics only specifies which processes can occur and which cannot. Physics theories aim to discover the operating rules of the universe, and everything should follow those rules. However, the intriguing thing is that thermodynamics tells us that even if the rules of the universe exist and are adhered to, the results may not be completely controllable. Some people describe the three laws of thermodynamics as a strange set of rules for a physics competition:

1. The law of conservation of energy: The competition is a zero-sum game.
2. The law of entropy increase: No matter how hard you play, you never have a chance to win. Also, as long as the game is long enough, you will probably lose.
3. The unattainability of absolute zero: Once you participate in the competition, you cannot stop competing.

Furthermore, there is an even more bizarre hidden rule of thermal equilibrium: Competitors with strong abilities must give points to those with weaker abilities until everyone is on an equal level. However, the absurdity lies in the unclear mechanism of distributing points, making it impossible to control the number of points given precisely. So, even though the intention is to have everyone possess the same ability, it can never truly be achieved, leading to an eternal cycle of reincarnation.

Thermodynamics is mainly summarized by the following three laws and the Zeroth law. Although the Zeroth law appeared last, it is considered more fundamental than the original three laws, thus earning the name "Zeroth law of thermodynamics."

First Law of Thermodynamics (Thermal energy and work): The total energy in a closed system isolated from the environment remains constant. The energy in the isolated system can only be transferred from one place to another or undergo a change in its form.

Second Law of Thermodynamics (Law of entropy): The entropy of a local region within a closed system may decrease, but the total entropy of the entire system will always tend to increase. Entropy is dependent on the state of the system but not on the process of how this state was reached.

Third Law of Thermodynamics (Entropy at absolute zero): The entropy of a thermodynamic system approaches a constant value as the temperature approaches absolute zero. For a perfect crystal, its entropy is zero at absolute zero. Another common statement is that it is impossible to bring the temperature of any thermodynamic system down to absolute zero by a finite number of operations. In other words, it would require an infinite number of steps to reach absolute zero, which means you will never get there.

The Zeroth Law of Thermodynamics (Thermal equilibrium): When two objects with different temperatures are placed in a closed system isolated from the environment, after a period of contact, the two objects will reach a state of thermal equilibrium with the same temperature.

In physics, there is a rigorous definition of the system and the environment. The system refers to the region where things happen, and the external space of the system is called the environment. The system boundary separates the system from the environment (Figure 4.1). The system boundary confines the system in a limited space, and matter and energy can be exchanged between the system and the environment through the system boundary. Suppose that a system changes from the original state to another new state through a specific process, and there exists a reverse process that can bring both the system and the environment to their original states. In that case, the process is called a reversible process.

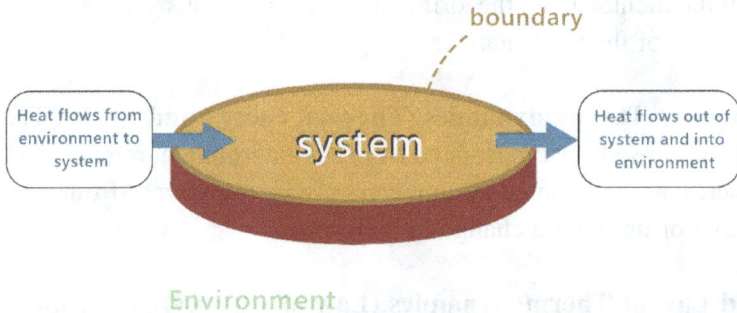

Figure 4.1. Schematic diagram of system, environment, and system boundary. An isolated system is a hypothetical system that has no exchange of energy and matter with the external environment.

By contrast, if no reverse process can make the system and the environment return to their original state, the process is an irreversible process. In everyday life, when a ball falls from a height and rebounds, its height gradually decreases over time until it stops on the ground. If we were to see the ball bouncing higher and higher on its own, we would think it was a miracle. Similarly, when high-temperature water and low-temperature ice are mixed together, they eventually reach a state of thermal equilibrium, which is also irreversible. The second law of thermodynamics states that in an isolated system, entropy will increase with the passage of time. The law implies time irreversibility, often referred to as the existence of an arrow of time. The directionality of nature's development also verifies the irreversibility of time, as exemplified by Darwin's theory of evolution. This reveals that time is something different from space — it possesses a directionality and irreversibility. Recently, time crystals have suddenly become a hot topic of discussion. A time crystal is an open system that maintains a non-equilibrium state with the surrounding environment, showing the characteristics of time-translation symmetry breaking. Time crystals were proposed by Frank Anthony Wilczek in 2012. Compared with normal crystals, which are arranged periodically in space, time crystals repeat periodically in time and exhibit a perpetual-motion behaviour.

An ideal system where only kinetic energy and potential energy are exchanged can return to its original state due to the law of conservation of energy. However, in real situations, the presence of energy dissipation leads to irreversible phenomena. For example, in an isolated system where energy is conserved, heat spontaneously transfers from an object with a higher temperature to an object with a lower temperature. The transition of an isolated system from an ordered state to a disordered state is an irreversible process, and nature always evolves in the direction of increasing disorder. The second law of thermodynamics introduces the increase of entropy to describe irreversible processes. The development of many macroscopic processes has directionality. Thus, entropy is often linked to the irreversibility of time, and the increase of entropy is used to define the direction of time flow ("arrow of time").

The idea of entropy is even more important in physics than the idea of energy. The first law of thermodynamics is the law of conservation of energy, which means that energy cannot be created or destroyed.

The second law of thermodynamics clearly states that even with energy conservation, processes must conform to thermodynamic principles: In a process from one equilibrium state to another equilibrium state, if the process is reversible, the entropy does not change; if the process is irreversible, the entropy increases. The law of entropy increase defines the direction of natural processes, with isolated systems inevitably developing spontaneously towards maximum disorder. In the three phases of water encountered in our daily lives — solid, liquid, and gas — their degrees of disorder vary. Therefore, they have different levels of entropy (Figure 4.2). Molecules in ice are confined to a solid structure, limiting their movement and thus resulting in low entropy. In contrast, molecules in water vapour can move almost freely, so water vapour has high entropy. In liquid water, molecules have more places to move to than those in ice but are not as free as in water vapour, so the entropy of liquid water is intermediate between ice and water vapour. When making ice cubes with water in a refrigerator, water and ice form a system, and the refrigerator is the environment in contact with the system. The entropy of the system decreases, causing water to turn into ice. However, since the entropy change during a process is related to the amount of energy transferred as heat, for the system's

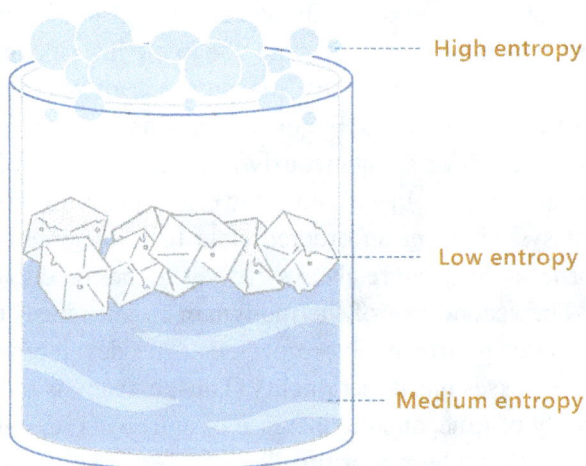

Figure 4.2.   Entropy differences of water, ice, and water vapour in a glass. Water molecules are least likely to change position in an ice cube, while water molecules in water vapour can move almost freely.

entropy to decrease, there must be heat flowing out of the system into the environment. In other words, the refrigerator does work to lower the entropy of water, allowing the formation of ice. More precisely, although the entropy of the system (water and ice) decreases, the increase in the entropy of the environment is greater than the decrease in the system's entropy. Therefore, the overall entropy of the environment and the system still increases. When the ice cubes are taken out of the refrigerator, the ice cubes will turn into water again. In this case, the room becomes the environment, and the entropy of the ice increases, absorbing heat from the surroundings and causing the temperature of the nearby environment to decrease.

4.2 Maxwell's demon and entropy

After Maxwell established the famous four equations of electromagnetism, his research interest turned to thermodynamics. Starting from the microscopic mechanism of molecular elastic collision, he derived the Maxwell–Boltzmann velocity distribution of macroscopic equilibrium state. After Maxwell, Boltzmann also provided a derivation of this distribution, in which he found that gas molecules tend to approach the Maxwell–Boltzmann distribution due to random collisions. Boltzmann argued that this tendency is the physical origin of irreversibility in the second law of thermodynamics. When two systems with unequal temperatures are brought into contact, fast-moving molecules will transfer energy to slow-moving molecules through collisions. Eventually, the two systems reach thermal equilibrium.

The law of entropy increase asserts that an isolated system will spontaneously develop in the most chaotic direction. According to the second law of thermodynamics, when the universe is regarded as an isolated system, the entropy of the universe will only increase over time. Therefore, the ultimate fate of the universe will be a state where entropy can no longer increase, and all processes that would lead to an increase in entropy will be unable to occur. This is the famous "heat death" hypothesis. The enigmatic second law of thermodynamics has constantly been challenged by many scientists. Although Boltzmann provided a microscopic explanation of the law, it still rests on specific assumptions.

Maxwell once proposed a thought experiment called "Maxwell's demon" to question the second law of thermodynamics. Maxwell considered an isolated system consisting of two chambers of gas with different temperatures. According to the Maxwell–Boltzmann distribution, there are high-speed and low-speed molecules in the high-temperature region and low-temperature region of the system, respectively. However, even in the high-temperature region, there exist low-speed molecules. And, there are a few high-speed molecules in the low-temperature region as well. As shown in Figure 4.3, Maxwell imagined a small demon controlling the gate of the channel between the high-temperature and low-temperature regions, allowing only high-speed molecules to move from low-temperature to high-temperature regions and low-speed molecules to move from high-temperature to low-temperature regions. Under the precise control of the small demon, the high-temperature region accumulates more high-speed molecules and the low-temperature region gathers more low-speed molecules. Without violating the law of conservation of energy, the demon's management causes the temperature in the high-temperature region to increase and the temperature in the low-temperature region to decrease.

In the absence of the small demon, from the macroscopic point of view, heat indeed transfers from high-temperature to low-temperature regions, but for individual molecules, high-speed molecules in the low-temperature region can still move to the high-temperature region. Therefore, Maxwell believed that the second law of thermodynamics is not like Newtonian mechanics, which can accurately describe the individual motion of molecules; instead, he believed it is a macroscopic statistical result arising from the collective behavior of a large number of molecules. We can already see the correlation between information and entropy here. As long as the small demon controlling the gate switch can accurately grasp the speed of incoming gas molecules, which is the information about the molecules, the entropy of the gas system can indeed be reduced.

The idea of Maxwell's demon clearly violates the second law of thermodynamics and is essentially an idea of a perpetual motion machine. Leo Szilárd (1898–1964) recognized that the problem with Maxwell's demon lies in "measurement." The purpose of measurement

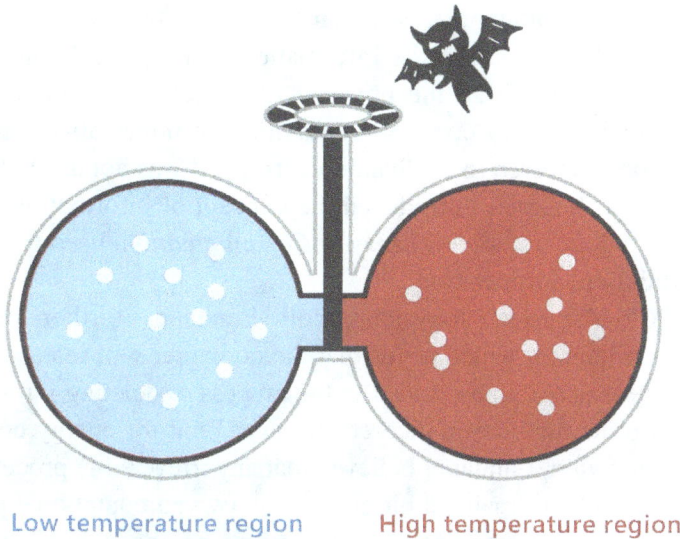

Low temperature region   High temperature region

Figure 4.3. Maxwell's thought experiment. The gas system is divided into a high-temperature region on the right and a low-temperature region on the left, with Maxwell's demon controlling the gate of the channel in the middle. The system, for probabilistic reasons, normally changes from an ordered, low-entropy state to a disordered, high-entropy state. This is because an ordered system has fewer microstates and thus a lower probability of occurring, while a disordered system has more microstates and thus a higher probability. However, Maxwell's demon seems to be able to challenge the second law of thermodynamics as long as it can coordinate the opening and closing of the gate in accordance with the speed of gas molecules.

is to gain information. In each instance of opening and closing the gate, there is a need for a binary bit of information to indicate whether it should be opened or closed. In 1929, building upon Maxwell's original idea, Szilárd proposed "single-molecule heat engines," each controlled by one small demon. These demons have to pay a price to obtain information, that is, the act of measurement will increase the entropy of the surrounding environment. Therefore, the decrease in the "thermodynamic entropy" of the gas system arises from the increase in the "information entropy" during the demon's measurement process. Even though the thermodynamic entropy of the system indeed decreases, the total of the information entropy and the thermodynamic entropy still obeys the second law of thermodynamics. Szilárd proposed two concepts that no one had previously mentioned, "information entropy" and "binary." He was the first to recognize the physical nature of information, linking information with energy consumption. Although information entropy is now commonly referred to as Shannon entropy, it was not until 1948 that Claude Elwood Shannon introduced the concept of information entropy in his work *A Mathematical Theory of Communication*, used to quantify the uncertainty of information.

In 1961, American physicist Rolf Landauer further proposed Landauer's principle, which states that the computer will release a small amount of heat to the environment in the process of deleting information. Landauer's principle gives a theoretical lower limit for energy consumption in computation. Landauer believed that any irreversible processing of information, such as erasing a bit or merging two computational paths, is accompanied by an increase in the entropy of the environment. By Landauer's principle, even if Maxwell's demon could determine the velocity of molecules without generating entropy (reversible measurement), it would inevitably have to erase the information of the previous molecule before measuring the next one, thereby leading to an increase in entropy. An intuitive view of Landauer's principle is that losing information from a system also means losing the ability to extract energy from the system. Since each bit has two possible states, 1 and 0, the entropy of a binary variable is $k_B \ln 2$, where the 2 in the natural logarithm represents the two possible states of the system, and k_B is the Boltzmann constant. Since the entropy change is related to the amount of heat transfer, the erasure of information inevitably involves the conversion of energy into heat, released from the system into the environment, which is why the computer keeps heating up. Every time a computer rewrites the information, it essentially turns the information into heat loss. According to Landauer's principle, erasing each bit of information at room temperature dissipates a minimum energy of 2.87×10^{-21} joules (~0.0175 eV) for an ideal two-level system. This might seem small, but if there are 10^{12} transistors changing states every second, the energy consumption rate will be about 3 kW. Moreover, in practice, the energy losses in a complementary metal-oxide-semiconductor (CMOS) or GPU system can even be up to millions of times greater than this number. Almost everyone is familiar with

Moore's law, which states that the number of transistors on an integrated circuit doubles every 18 months, leading to a corresponding doubling of performance. However, along with the progress of Moore's law, the number of bits within the same area also increases. According to the second law of thermodynamics and Landauer's principle, the more the number of bits, the greater the energy consumption. Presently, Bitcoin mining machines and large-scale data centres are extremely energy-intensive due to the continuous transformation of vast amounts of information into heat that is lost to the environment. The only solution to this substantial energy consumption issue is to implement reversible computing, where the energy of information does not dissipate as useless heat into the environment.

4.3 Significance and influence of information entropy

4.3.1 *Information entropy*

In physics, matter and energy are quantified, and their mathematical models are established. A quantified description of information is also necessary to clarify the physical nature of information and establish the foundational theory of information science. Can information be quantified? Clearly, yes. The passage you are currently reading may have several hundred words, while each chapter of this book contains at least thousands of words, and the entire book consists of over 150,000 words. It is evident that there is a significant difference in the amount of information conveyed among these three cases. But, can information be accurately and effectively quantified solely based on the number of words? When using different languages, the amount of information conveyed by the same number of words is naturally different due to differences in cultural background. Even if using the same language, people with varying levels of proficiency may convey vastly different amounts of information with the same word counting. Can we quantify the amount of information in a file by its size? The amount of information in various formats of files is also very different. A PDF format often contains much more information than a Word document for files of the same size.

Claude Shannon is known as the "father of information theory" because he provided a quantification of information and its scientific meaning. In Shannon's definition, the measure of information is a measure of uncertainty, as information serves to eliminate uncertainty. The uncertainty in physics corresponds to the degree of disorder in an isolated system, which is the entropy mentioned in Section 4.1. Once enough information is available, the system's uncertainty decreases, reducing the degree of disorder in a thermodynamic system. Information can be thought of as negative entropy, decreasing the disorder of a system. Based on thermodynamic principles, reducing the level of disorder of a system requires applying energy to do work on the system. Hence, the connection between information and energy is evident. In 1948, Shannon introduced entropy into computer science as a measure representing information content, known as Shannon entropy. It was said that John von Neumann suggested that Shannon should name the concept of eliminating uncertainty "entropy," mainly because this enigmatic term is likely to impress those unfamiliar with the field.

Information science is a significant part of computer science. Modern computer technology has begun to develop rapidly because of the effective quantification of information. Nowadays, the concept of information entropy is commonly used to quantify the amount of information. But, how do we measure the "information content" of an event? How much information can we get from a single frame of a TV show? How much information can be absorbed from reading this book? As previously mentioned, the quantification of information is directly related to the uncertainty of events. The information content of an event signifies the magnitude of various potential variations of the event. The information content is smaller when the event is more definite. By contrast, the information content is larger when the event is more ambiguous. For example, the statement "I am traveling from Shanghai to Beijing" provides very specific information about the destination; hence, we are 100% certain that the destination is Beijing, so the information content regarding the destination is zero. However, it is not clear when to go, how to go, why to go, and who is traveling along, leading to various possibilities. Thus, the information content about other aspects, such as timing and transportation, is large. If the statement contains a little more description, such as

"I am traveling with my son from Shanghai to Beijing" or "I am taking a flight from Shanghai to Beijing," the information content regarding the destination remains unchanged, as going to Beijing remains a definite fact. If we possess adequate understanding of an event, we do not need much additional information to grasp the situation. However, if an event is highly uncertain, a substantial amount of information is necessary to know what will happen. Therefore, uncertainty can serve as a quantitative measure of information content.

"Information content" is a very abstract thing. We often say that there is a lot of information content or less information content, but it is difficult to precisely quantify how much information content there is. It was only after Shannon that the quantification measure of information content was established. The amount of information content is related to uncertainty or probability. The information content of an event is a function of the probability of the event. The information content of an event was defined by Shannon as $-\log_2 P(x)$, where $P(x)$ is the probability of the event occurring. The smaller the probability of the event, the larger the information content of the event. When the uncertainty of an event is higher, it becomes less predictable, indicating greater information content.

"Entropy" was originally from thermodynamics and used to measure the degree of disorder or uncertainty in a system. Today, the concept of "entropy" has been widely applied in science, mathematics, management, and other fields. How is "information entropy" rigorously defined? Higher information entropy corresponds to greater disorder in a system, indicating larger information content. This seems counter-intuitive — why does a more ordered state not have greater information content? Let us illustrate information entropy with a simple example. The commonly used unit of "information entropy" is typically the "bit." Consider a toss of an ideal (fair) coin, which can result in heads or tails. The probability of getting head (0) and the probability of getting tail (1) are equal, both one-half. The information entropy of this event (a coin toss) is called one "bit." If n independent tosses are carried out, the information entropy is n "bit," because it can be represented by a bitstream of length n (See figure 4.4 for an example of $n = 2$). However, if there is a biased coin with identical sides, the result of tosses will always be heads. Since the outcome of the tosses can be precisely predicted in this case, the probability $P(x)$ is 1 and

$$H = -\Sigma_i P_i \, log_b P_i$$

When b = 2

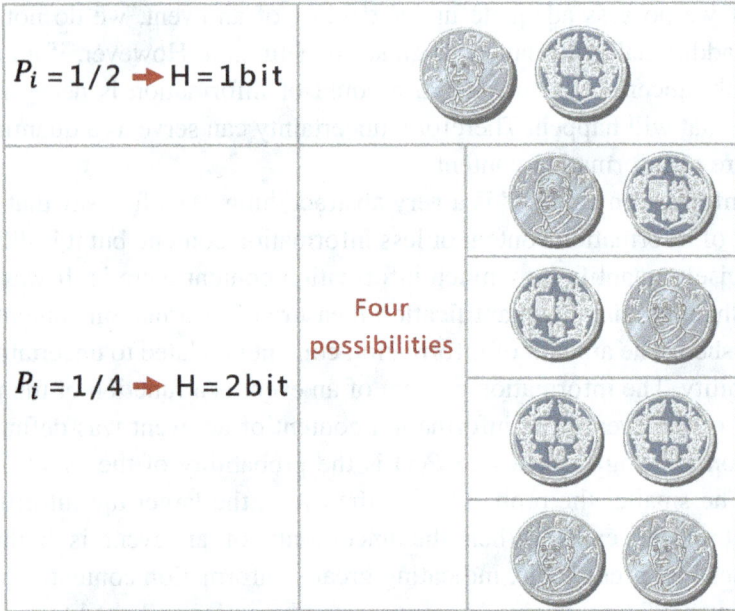

$P_i = 1/2 \rightarrow H = 1\,bit$

Four possibilities

$P_i = 1/4 \rightarrow H = 2\,bit$

Figure 4.4. Shannon's information entropy (*H*) is the average information content of all possible outcomes of variables, given by $H = -\Sigma_i P_i log_b P_i$, while Boltzmann's thermodynamic entropy is represented as $S = -k_B \Sigma_i P_i log_e P_i$. Information entropy (*H*) and thermodynamic entropy (*S*) are proportionally related; the greater the disorder in a system, the larger *S* and *H* become. When the base of the logarithm *b* = 2, the unit of information entropy is "bit"; when *b* = *e*, the unit is "nat"; and when *b* = 10, the unit is "Hart." When information entropy is in bits, the bit count is the information entropy. For example, a coin toss has two possible outcomes, resulting in information entropy of one bit; a toss of two coins has four possible outcomes, resulting in information entropy of two bits.

the information entropy is zero. For a system that can be inferred exactly, that is, a completely ordered system, the information content is zero.

Let us further consider a slightly more complicated example. Suppose there are three buckets, each with four balls. In bucket A, there are four red balls. In bucket B, there are three red balls and one blue ball. In bucket C, there are two red balls and two blue balls. If we randomly draw a ball from the buckets, from bucket A, we will definitely draw a red ball.

In bucket B, there is a 75% chance of drawing a red ball and a 25% chance of drawing a blue ball. In bucket C, there is a 50% chance of drawing either a red or a blue ball. Therefore, bucket A is in a state of complete certainty, bucket B has some degree of uncertainty, and bucket C has the highest level of uncertainty. As discussed earlier, the higher the degree of disorder and uncertainty, the greater the value of information entropy. If the balls in a bucket have many possible rearrangements, then the bucket has high entropy, and if there are few possible rearrangements, then the bucket has low entropy. The number of possible rearrangements of balls can be counted: Bucket A has only one arrangement, bucket B has four possible positions for the blue ball, and bucket C is the most complex, with six possible arrangements for the blue balls. In terms of intuition, a new blank USB drive is like bucket A. It has been formatted and is very ordered. After a long period of use, it becomes like bucket C with high disorder and large information entropy.

Shannon's information entropy, denoted as H, is the average information content of all possible outcomes of variables, given by $H = -\Sigma_i P_i \log_2 P_i$. Information entropy measures the additional information needed, on average, to determine whether a drawn ball from a bucket is red or blue. For bucket A with all red balls, no additional information is required, so the information entropy is naturally 0. In the case of bucket B, the probability of drawing a red ball is 3/4, which makes the $\log_2 P_i$ of red balls -0.415. For the blue ball, with probability P_i being 1/4, the $\log_2 P_i$ is -1.998. Since there is a 3/4 chance of drawing a red ball and a 1/4 chance of drawing a blue ball from bucket B, the information entropy is calculated as $(0.415 \times 3/4) + (1.998 \times 1/4) = 0.8113$. In bucket C, with an equal number of red and blue balls, the probabilities of drawing a red ball and a blue ball are both 1/2. Therefore, $\log_2 P_i$ is -1 for both possibilities. Accordingly, the information entropy of drawing a ball from bucket C is calculated by multiplying the individual information content of drawing a red ball and a blue ball by the respective probabilities (2/4 for red and 2/4 for blue) and taking the sum, which results in 1. If there is another bucket D with three blue balls and one red ball, and another bucket E with four blue balls, the information entropy of drawing a ball from these buckets can be visualized in Figure 4.5(b). Shannon's information entropy is closely related to Boltzmann's thermodynamic entropy. For example, when the volume of

(a)

(b)

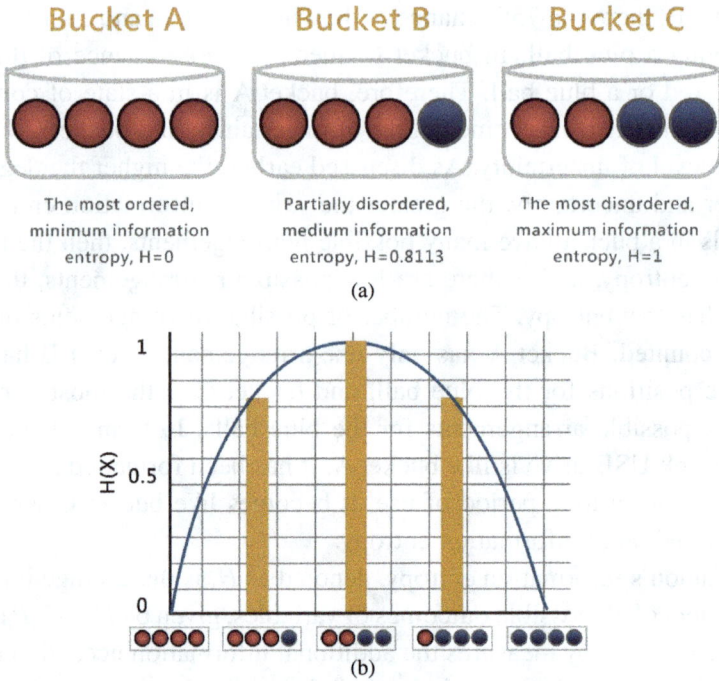

Figure 4.5. (a) Information entropy is directly related to the degree of disorder of the system. (b) The relationship between information entropy of drawing a ball and the combinations of coloured balls in each bucket.

an ideal gas is compressed by half at a constant temperature, the thermo-dynamic entropy decreases ($\Delta S < 0$). In this process, the uncertainty of the molecules' positions decreases due to the reduced spatial distribution of molecules (Figure 4.6). Therefore, the information entropy also decreases ($\Delta H = \Delta S$).

4.4 Reversible quantum computers

4.4.1 *Reversible computing*

A "Maxwell's demon" that does not consume energy does not exist. The little demon must acquire "information" to decide whether to open or close the gate, and obtaining this "information" requires the expenditure

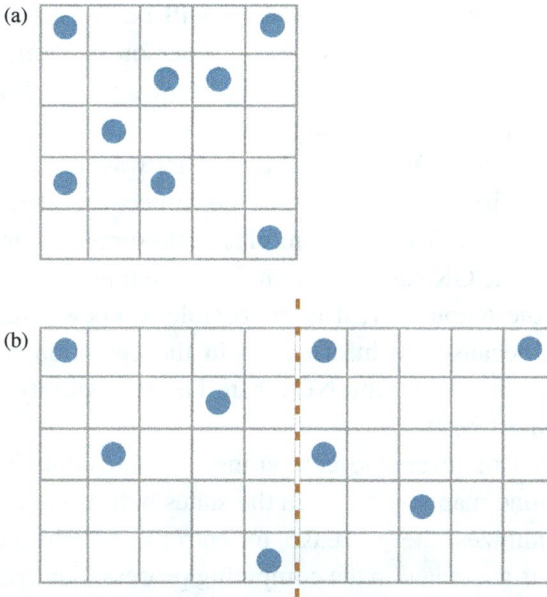

Figure 4.6. (a): a region with a smaller volume. (b): a region twice as large as shown in (a). The circles represent gas particles. When the same number of gas molecules is distributed in region (a) and region (b), the possible arrangements in region (b) are significantly greater than those in region (a). In an ideal gas with no intermolecular forces, during isothermal compression from (b) to (a), the ways in which gas molecules can be arranged decrease, resulting in a reduction in entropy.

of energy. The thought experiment has led to a new understanding of the world: In addition to matter and energy, there is information. A. G. Oettinger from Harvard University once said, "Without matter, nothing exists; without energy, nothing happens; without information, nothing makes sense." Landauer also put forth another intriguing argument: "Forgetting," that is, erasing information, also consumes energy, and forgetting is logically irreversible. In order to save energy, Landauer came up with the concept of "Reversible computing" and conducted research with his colleague Charles L. Bennett. Reversible computing aims to reduce the energy consumption of computers by retaining all information throughout the computation process.

To understand reversible computing, one must first know what irreversible computing is. For example, if you open the calculator on your mobile phone and press 3 + 2, the result will be 5. This 5 appears as a result of your input of 3 + 2. However, if a person suddenly shows you a 5 on their calculator, you cannot possibly know how this 5 was obtained. This five may come from inputting 3 + 2, 1 + 4, or a complex equation like (268 + 152)/20−16. We cannot know what was input because digital calculations are irreversible, where the preceding data are erased. Similarly, most of the logic gates in classical computers are irreversible operations. Take the OR gate, for example; if the input is 10, 01, or 11, the output is 1. If the output is 1, it is impossible to know what the original input value is, because the information in the operation process is discarded. On the other hand, the NOT gate has reversibility; if the output is 1, the input must be 0.

In the model of reversible computing, the transition function establishes a one-to-one mapping between the states before and after the transition. This minimizes the increase in entropy, which means that no additional heat is generated in the computing process. For irreversible logic gates, the input state is lost after operation, resulting in output information being less than input information. According to Landauer's principle, the loss of information will be dissipated into the environment in the form of heat. Reversible logic gates, on the other hand, transform the input computational states to the output computational states without any loss of information, thus not consuming energy. If the logic gate is used as a model to discuss reversible computation, as long as the mapping between the output and the input is not one-to-one, it is irreversible computation. The amount of information decreases after operations involving irreversible logic gates, such as OR gate or XOR gate. Any reversible logic gate needs to have the same number of input and output terminals. There are two single-input reversible logic gates: the NOT gate and the YES gate. For logic gates with two input terminals, the reversible logic gate is the Controlled-NOT gate (CNOT Gate). It treats the first input as a control reference and performs a NOT operation on the second input if the first input is 1. As the first bit, the control bit, is kept unchanged, the CNOT gate reversibly computes the XOR function and outputs the result on the second bit. The three-input Controlled-Controlled-NOT gate, also known as the Toffoli gate, was proposed by Tommaso Toffoli. It is a universal reversible logic gate with three inputs and three outputs. If the first two inputs are 1, it will invert the third input; otherwise, all inputs remain unchanged. The Toffoli gate can be used to implement any Boolean function on a classical computer. In addition to the output bit, reversible logic gates require many extra bits to store the history of operations. In ideal situations, the entropy remains unchanged during the process of reversible computation.

In quantum computing, the unitarity of quantum systems inherently guarantees the reversibility of computations. Since quantum computations require reversibility, the quantum logic gates must be reversible. As mentioned earlier, the Toffoli gate is reversible and universal for classical computing. Therefore, the quantum Toffoli gate can be implemented on quantum computers, which means that quantum computers can perform all logic operations of classical computers. Yaoyun Shi discovered that a set of universal quantum logic gates can be constructed using only the three-qubit Toffoli gate and the Hadamard gate on a single qubit, allowing for the generation of arbitrary quantum operations. This implies that quantum computation goes beyond classical computation by adding the Hadamard gate. To put it another way, quantum computation is classical computation enhanced with the Hadamard gate. In classical circuits of classical computing, using an NAND (or NOR) gate alone can generate functions of all logic gates, whereas in quantum circuits, at least a Toffoli gate and a Hadamard gate are required.

In quantum computing, much like the depicted reversible state of multiple spaces in the movie *Tenet*, quantum operations take place in the high-dimensional Hilbert space spanned by multiple qubits. The crucial aspect in quantum computation is that since no information is deleted during the process, the energy loss concerns addressed by Shannon and Landauer do not arise. This advantage allows for continuous reversible computations in a high-dimensional mathematical space in a reciprocating manner. However, Hilbert space is an abstract mathematical space, which is completely different from the physical space in the movie *Tenet*. The energy losses inherent in the real world make the notion of resetting life events portrayed in science fiction movies impossible.

4.5 Concluding remarks

Without information, there would be no modern computer science. Since Boltzmann proposed entropy, information has gradually been associated with physics through the connection of energy. Can fundamental concepts in physics, such as causality, irreversibility, disorder, and chaos, be newly understood on the basis of information? In quantum mechanics, whether entanglement and non-locality are related to information remains to be explored. The foundation of reversible computing is the conservation of information. Because of the conservation of information, there is no energy loss. Matter and energy were once believed to be two independent and unrelated elements until Einstein discovered that they are inter-changeable, relating them through the famous equation $E = mc^2$. Following the work of Szilárd and Landauer, information was also found to have a certain relationship with entropy and energy. But, our current understanding of the relationship between information, matter, and energy is still vague. In the agricultural age, human beings began to study matter because they needed tools for cultivation. After the Industrial Revolution, humanity's increased need for energy led to the discovery of various forms of energy. This revolution deepened our understanding of energy and, consequently, led to the realization of the close relationship between matter and energy. The ongoing information revolution has brought to light the potential connection between information and energy. We now understand that changes in entropy imply alterations in information. Perhaps information may also be associated with matter through the connection between information and energy. The intricate relations between these elements remain a subject for future research in order to unravel the mysteries of information. In the microscopic world, mass and energy are interchangeable; in the macroscopic world, thermodynamic entropy is closely related to information entropy, and deleting or acquiring information consumes energy. Currently, it is quite clear that there exists a triangular relationship between matter, energy, and information. Matter and energy can be connected through Einstein's $E = mc^2$, and the relationship between energy and information can be linked through Landauer's theory and Boltzmann's theory. But, is there a direct bridge to relate information and matter? The connection between matter and energy is established through the laws of the quantum world and relativity, while the connection between information and energy arises from macroscopic thermodynamics and statistical mechanics. Even though both relations involve energy, whether energy can truly serve as a direct bridge between macroscopic and microscopic worlds remains a question due to discrepancies between the laws of the microscopic and macroscopic worlds. Or, is it only meaningful at the so-called mesoscopic scale, which lies between the microscopic and macroscopic world? Is there reincarnation in life, and is computation reversible? How do quantum probability waves collapse into classical measurement results? How are the microscopic, mesoscopic, and macroscopic worlds connected? Entropy is a macroscopic property and is related to information, but which is more important among information, energy, and matter? The mystery is awaiting further research. In the future, physics, information science, computer science, cognitive science, and biology may potentially inspire more astounding knowledge within this interdisciplinary field, which is highly anticipated.

Chapter 5

Qubits and Quantum Computers

Middle beads, in the abacus; Upper beads, above the backbone, also on the left; Lower beads, below the backbone, also on the right; Backbond, the beams and partitions in the abacus. During digital carry operation, shift the lower beads to the upper beads.

— Song, Xie Chawei (~1000)

Inside the Schrödinger's cat box, the quasi quantum particles are dancing on the net of quantum attention function, vanishing and arising.

— Amit Ray (1960–)

5.1 Quantum properties

5.1.1 *Introduction*

At present, the semiconductor manufacturing process has approached the atomic limit of a few nanometres. Quantum properties have become increasingly important because of finite size effects, and bottlenecks caused by various physical limitations have gradually appeared, making the mass production process of semiconductors more and more expensive and energy-intensive. In recent decades, quantum-related technologies have developed rapidly. The precision of these technologies has become increasingly higher, and it is even possible to manipulate atoms to simulate quantum systems. Multinational companies such as Google, IBM, Intel, and Microsoft have expanded their quantum technology R&D teams. Developed countries have also invested billions or even tens of billions in extending quantum applications. The development direction of the qubit is also very diverse. Superconductors, ion traps, cold atoms, nitrogen-vacancy centres in Diamond, quantum dots, Silicon-based Qubits, topological qubits, photonic integrated circuits, and nuclear magnetic resonance have all been used in different types of quantum computers. An entire ecosystem of the quantum industry is gradually taking shape, with numerous start-ups springing up. Companies in other fields have also begun to merge or cooperate with quantum start-ups to develop new forms of quantum applications. The quantum industry has entered the era of regional hegemony, and the battle for hegemony has emerged in Quantum Valley after the Silicon Valley era.

After Google demonstrated quantum supremacy in 2019, the word "quantum" has frequently appeared in media pages around the world and has become a buzzword for modern people. However, quantum mechanics is still unfamiliar and intimidating to ordinary people. Even many scientists think quantum computing is mysterious, difficult to understand, and has little chance of success. Even though scientists have proposed that the UN name 2025 the "International Year of Quantum Science and Technology," it is still unsure when exactly the Year to quantum (Y2Q) will arrive or if it will not appear at all like the Y2K bug. This chapter will answer two questions: What is the nature of quantum? How is quantum computing different from classical computing? Let us slowly uncover the mystery of quantum computing for readers, introduce where the mighty computing power of quantum computers comes from, and, most importantly, discuss the future of quantum computers.

5.1.2 *What is quantum?*

Quantum is indeed an esoteric term for the general public. In addition, because quantum phenomena do not appear in daily life, such terms are often used by people with intentions to fool the public, resulting in such ridiculous nouns and myths such as quantum water, quantum underwear, pills tunnelling through bottles, and retrieving medicine in the air. The concept of quantum is the most important difference between modern physics and classical physics. It was first put forward by German physicist Max Planck. The word quantum comes from the Latin "Quantus," which means "how much." It is the smallest indivisible basic unit in the physical world. If a physical quantity has the smallest indivisible unit, then it is called a quantum. In the classical world, physical quantities change continuously, such as temperature, height, and weight. Unfamiliarity with discontinuous changes is the main reason why everyone is confused about quantum phenomena. Imagine you are shopping in a supermarket. There is the smallest unit for the amount you need to pay at the checkout, e.g., "cent" in the US. Any item in the supermarket will be priced in the minimal unit "cents," and the price will vary in a discrete manner between different items. There will never be a price with a decimal point of "cent." This is the best analogy for the discontinuous phenomenon in the quantum world. In the microscopic world, physical quantities such as energy and momentum cannot be infinitely divided into infinitesimals, but have a minimum basic unit. These physical quantities are said to be "quantized" in the microscopic world. In China, Mozi also proposed that when matter is divided to a certain extent, it cannot be divided further. In the book of Mozi, it is mentioned that "empty" and "non-half" cannot be split. It is also mentioned in Buddhist scriptures that something can be divided if it is not a basic unit, but it cannot be divided if it is a basic unit (非極微復有餘極微，是故極微非有分). The non-half idea in Mozi and basic unit in Buddhist are very similar to the idea of quantum in the microscopic world.

5.1.3 *Quantum properties*

Because we live in a macroscopic world, whether in an Eastern or Western cultural environment, there is no life experience related to quantum phenomena in our daily life. This "quantum alienation" leads to the absence of quantum concepts in human intuition. Therefore, when quantum technology suddenly starts to be integrated into our daily life, it causes a great knowledge shock. The following will introduce the representative and counter-intuitive quantum principles and phenomena of quantum superposition, quantum entanglement, and quantum measurement:

1. **Quantum superposition:** In our experience, objects can only be in a specific state of time and space. However, particles in the microscopic world are in a linear combination of multiple states, called "superposition states." A particle may be in one of these multiple states, and will not become a definite state until it is observed. Like a coin rotating in the air, the process of continuous rotation is similar to a superposition state composed of a head state and a tail state. It is not possible to determine whether the result is heads or tails until the rotating coin stops on the ground. This situation is what we often hear about as "Schrödinger's cat." Before the box is opened, the state of life or death is completely uncertain. This conflicts with our daily experience that "things that exist always have a definite state." Whether the chandelier in the living room is on or off, there can be no ambiguous answer. But, in the quantum world, objects can indeed be in a superposition state. An atomic-sized light bulb can be both on and off at the same time. This strange property in the microscopic world is one of the main sources of the power of quantum computing. To be, or not to be, that is the question — this is a big question for Hamlet in William Shakespeare's screenplay, but it will never be a question in the microscopic world.

2. **Quantum entanglement:** In quantum mechanics, after a group of particles interacts, it can form an overall state in which it is no longer possible to describe the individual states of the particles independently. This phenomenon is known as quantum entanglement. The inseparable state is called an entangled state. It is the most bizarre phenomenon in quantum mechanics. Taking the two-particle entangled state as an example, the states of the two particles in an entangled state cannot be determined until they are measured. However, no matter how far apart they are, as long as the entangled state is not destroyed, once one particle is measured, the state of the other particle will be determined. Einstein referred to this long-range correlation as "spooky action at a distance." The inseparability of an entangled state

is like the poem *Married Love* written by Kuan Tao-sheng in the Yuan Dynasty: "Take a piece of clay, twist a you, and mold a me. Break the two of us together, mix them with water, twist a you again, and mold a me. You are in my mud, and I am in your mud. In life we share a single quilt and in death we will still share a single coffin." Quantum entanglement is an essential resource for realizing quantum communication. And, because it is susceptible to environmental changes, it can also be utilized to make very precise and sensitive quantum sensors.

3. **Quantum measurement:** Quantum measurement is the core issue of quantum mechanics. Quantum measurement is different from the measurement in classical mechanics, as quantum measurement has an effect on the measured system and changes the state of the measured system. The weirdness of quantum measurement is best demonstrated by the single-electron double-slit interference experiment, in which measuring at the slit will make the interference pattern disappear. The act of measurement causes the loss of coherence, the definite phase relationship of the waves at both slits, resulting in the loss of the wave feature of electrons. This result also brings out an important notion of "quantum decoherence." In addition to measurements that will affect quantum systems, quantum systems also lose their quantum features and return to classical behaviour due to their inevitable interaction with the environment. This process is called quantum decoherence. The coherence of quantum states in quantum computers must be preserved to perform quantum computation. All kinds of hardware design efforts for quantum computers are mainly aimed at prolonging coherence time. Progress over the past decade has been significant.

5.2 Quantum turing machine

5.2.1 *Turing machine and classical computer*

The oldest known calculator is the abacus that has been around for a long time. It is unclear which individual created the abacus; however, it seems to have been created by the ancient Sumerian in Mesopotamia between 2700 BC and 2300 BC. The earliest known Chinese abacus dates to the

2nd century BC. More than 300 years ago, French mathematician Blaise Pascal invented the mechanical adder using the principle of gear drive. One cannot talk about modern computers without mentioning Alan Turing. Classical computers have evolved from the giant electronic calculators that occupied entire floors 40 years ago to laptops and smartphones that people can put into their pockets. Every one of these devices is essentially a Turing machine. The speed of modern computers has increased dramatically, and they are capable of performing a variety of unimaginable tasks. Without the concept of "effectively computability" proposed by Turing, computers might still exist today, but they would certainly look different. The British mathematician stated in his work "On computable numbers, with an application to the Entscheidungs problem," published in 1936, that all algorithms and functions that are effectively calculable can be computed by a Turing machine. That is to say, a programme written in any programming language can be executed by computing hardware. This statement is also known as the Church–Turing thesis. It combined abstract computability theory with actual computing hardware, and also laid the foundation for the development of modern computer science. Since "functions that are effectively calculable" are not precisely defined, the Church–Turing thesis cannot be formally proved mathematically. However, the Turing machine did provide the theoretical model for modern computers, and the information revolution thrived on vague heuristic definitions. There are still many problems that cannot be effectively solved by a Turing machine; as a result, finding a new model of computation that is better than Turing machines has been the Holy Grail for the information community.

The basic unit of information science is called a bit, which is composed of, stored in, calculated by, and transmitted by 0 and 1 in binary. The simplest way to implement bits is to use a two-state physical system. For example, the switching of electron tubes, light pulses in optical fibres, and magnetization in magnetic tapes can all serve as a two-state system. Today's computer chips use transistors as switches, quickly switching between 0 and 1 to store digital signals and perform various logical operations. Classical computers can only carry out fast-sequence operations and obtain a definite result of 0 or 1 after digital calculation, similar to using electronic properties to quickly manipulate the beads of a giant abacus.

Classical computers are always in a state consisting of 0s or 1s, but classical physics is continuous. Therefore, it is insufficient to use a Turing machine to simulate the classical physical world. On the other hand, since we already know that quantum mechanics describes our world more completely than classical mechanics, quantum computers whose operations harness quantum properties are more suitable for simulating nature. In 1983, Richard Feynman proposed to use a quantum simulator to realize quantum computing for the first time, and subsequently Paul Benioff designed a quantum mechanical model of the Turing machine to prove that Feynman's idea is indeed feasible. In 1985, David Deutsch proposed the "Church–Turing–Deutsch Principle," which he called "the physical version of the Church–Turing thesis," stating that every finitely realizable physical system can be perfectly simulated by a universal computing device. Deutsch proposed that a universal quantum computer, also called a quantum Turing machine, is such a device. He also proposed the quantum circuit model of computation, the most widely used model of quantum computation. In 1993, Professor Chi-Chih Yao of Tsinghua University developed a complexity theory of quantum circuits and showed that the quantum circuit model is equivalent to the model of quantum Turing machines. His work established the basis of current quantum computers.

5.2.2 Quantum turing machine

Deutsch's quantum Turing machine (QTM) and Turing's probabilistic Turing machine (PTM) have similar definitions, except the probabilistic nature of the former directly comes from the property of quantum mechanics. There is almost no difference in structure. The real difference is that the quantum state is represented by the orthogonal basis of the multi-dimensional Hilbert space, so the functions in the QTM must be quantized. But because the Turing machine itself is discrete, the process of quantization is concretely feasible. In QTM, the evolution of the quantum system is achieved through a unitary matrix, and the transition of the state is performed continuously, starting from the initial state till the terminal state is reached. The advantage of QTM over PTM is as follows: (1) In PTM, measurement is performed after each operation, so all information is lost during the process. However, in QTM, quantum measurement is only performed after all operations are completed, and all combinations and information during the process are still retained. (2) In QTM, quantum probability (amplitude) can undergo destructive or constructive interference. Therefore, as long as the instructions of the quantum algorithm are properly designed, the correct result can be amplified and directly observed, which is very difficult to achieve in PTM.

5.2.3 *Origin of quantum computers*

For the development of quantum computing, perhaps the most important question is the one posed by Feynman in 1981: Can classical computing effectively simulate quantum systems? If we intend to solve problems of macroscopic physics, classical computation is very effective. The changes in a classical physical system at each moment can be calculated through the equations of motion of the system. More importantly, classical physical systems usually have finite variables and a well-defined sequence of events. In many fields, even with discrete approximations, satisfactory results can be obtained using supercomputers. However, supercomputers become inadequate when it comes to solving problems in the microscopic world. In the microscopic world, the state of a system is described by a wave function. Therefore, what we need to deal with is the equation of motion of the wave function. As long as the wave function is known, all the properties of the system can be obtained. But, it is a huge task to describe the wave function in chemical molecules accurately. The chemical properties of the molecule depend on the collective behaviour of all the electrons within it. Each electron exists in the form of superposition of multiple states. Moreover, the quantum state of each electron is correlated with the quantum states of other electrons. Therefore, the problem becomes multi-dimensional, which has numerous variables, requiring almost infinitely large computing resources. Even the electron wave function in very simple molecules is beyond the reach of supercomputers. Feynman therefore stated that "... because nature isn't classical, dammit, and if you want to make a simulation of nature, you'd better make it quantum mechanical." Computers based on classical bits cannot be effectively extended to high-dimensional Hilbert space to deal with quantum problems. We cannot use classical computers to simulate nature.

Since supercomputers constructed with classical bits cannot simulate quantum systems, the natural idea is to use known quantum systems for this purpose. The approach is to use components with quantum properties of superposition and entanglement to build an effective system that can be manipulated and conforms to quantum principles, and then simulate quantum systems of interest on this system. The model of quantum computation states that processes in the microscopic world can be simulated by a quantum computer in principle. To do that, quantum information encoded in quantum states is manipulated in the Hilbert space using mathematical operations of linear algebra, which are called "quantum gates," in a quantum computer. Quantum gates are reversible unitary transformations operating on qubits in a quantum computer. Using various quantum gates, we can construct a quantum circuit that meets our needs, and then measure the final quantum state and output the result. Through quantum circuit design, programmers can combine quantum gates into various quantum algorithms to solve diverse problems. In Chapter 4, we stated that reversible computing is different from classical computing. Quantum gates correspond to unitary matrices in mathematics, which are invertible matrices, meaning the operation process itself has no energy consumption. It opens up the possibility of reversible computing using quantum theory. However, it is very challenging to realize multi-particle entanglement in practice. And, the entanglement is also susceptible to environmental noise. Currently, a fault-tolerant universal quantum computer is still a long way off and requires more effort to realize. The difficulty of physically implementing a quantum computer is very similar to the acrobatic performance of "plate spinning," in which the performer uses bamboo poles to spin multiple plates at great heights simultaneously. And, in addition to ensuring that each plate does not fall, it is also required that each plate spin at the same rate. Quantum computers are currently in the Noisy Intermediate-Scale Quantum (NISQ) era, where they can only spin a few plates at the same time. But, to realize a universal quantum computer, at least millions of plates must be spun simultaneously.

5.3 Quantum bit (Qubit)

In everyday life, there must be definite descriptions of things. For example, a light bulb can only be on or off, not on and off at the same time. The same is true for the bits in classical computers, which are either 0 or 1. Each classical bit can only store one piece of information. On the other hand, in quantum computers, the state of a qubit is not determined before measurement. Quantum superposition states can be any probability combination of 0 and 1, which is a capability that classical computers are unable to achieve. In mathematics, it is customary to use the Dirac notation to express the state of a qubit: $|\psi\rangle$ (pronounced "ket psai"). If a and b are the components of $|\psi\rangle$ on the unit vector $|0\rangle$ and $|1\rangle$ (pronounced "ket 0" and "ket 1"), respectively, the quantum state is represented as $|\psi\rangle = a|0\rangle + b|1\rangle$. After measurement, the probability of the state being in $|0\rangle$ is given by the modular square $|a|^2$, the probability of the state being in $|1\rangle$ is given by the modular square $|b|^2$, and the sum of the two modular squares is 1. For example, if the state of a qubit is $|\psi\rangle = 0.8|0\rangle + 0.6|1\rangle$, the probability of this state being in $|0\rangle$ is $|a|^2 = |0.8|^2 = 0.64$, and the probability of this state being in $|1\rangle$ is $|b|^2 = |0.6|^2 = 0.36$. The sum of the two probability must be 1. Quantum states can also be represented by vectors as linear combinations of states: The state of a qubit $|\psi\rangle = \begin{bmatrix} a \\ b \end{bmatrix} = a\begin{bmatrix} 1 \\ 0 \end{bmatrix} + b\begin{bmatrix} 0 \\ 1 \end{bmatrix}$ where $|0\rangle = \begin{bmatrix} 1 \\ 0 \end{bmatrix}$ and $|1\rangle = \begin{bmatrix} 0 \\ 1 \end{bmatrix}$.

It must be remembered that $|a|^2 + |b|^2$ is always equal to one.

All possible states of a qubit are often represented geometrically by a Bloch sphere. A classical bit only exists in two states: the north pole corresponding to "1" and the south pole corresponding to "0." The state of a qubit, in addition to the north and south poles, can exist in an infinite number of states distributed across the Bloch sphere. Each state corresponds to a point on the surface of the Bloch sphere, which has a radius of 1, and points at different latitudes will have superposition states composed of $|0\rangle$ and $|1\rangle$ in different proportions. In addition to the probability amplitude, the state of a qubit also has a phase factor, which corresponds to longitudes on the Bloch sphere. The Bloch sphere helps us visualize quantum gate operations on a single qubit. Performing an operation on a qubit changes one quantum state to another. In Bloch sphere representation, this corresponds to transforming a point on the surface of the sphere into another point. The transformation is a unitary transformation corresponding to rotations of the Bloch sphere. Performing a series of operations on qubits is equivalent to performing a series of unitary transformations. These operations are reversible, meaning the computation process consumes zero energy.

The Bloch sphere is a helpful visualization of the quantum state of a single qubit. But, the high-dimensional space expanded by multiple qubits cannot be easily visualized in this way. If we have two qubits, and each qubit has the states of $|0\rangle$ and $|1\rangle$ simultaneously, as a whole, it will correspond to the four states of $|00\rangle$, $|01\rangle$, $|10\rangle$, and $|11\rangle$. All the possibilities of the superposition state composed of these four states far exceed the combination of two separate two-dimensional spheres as shown in Figure 5.1. Similarly, if there are N qubits, 2^N states can be represented at the same time. The ability of qubits to expand computational space at an

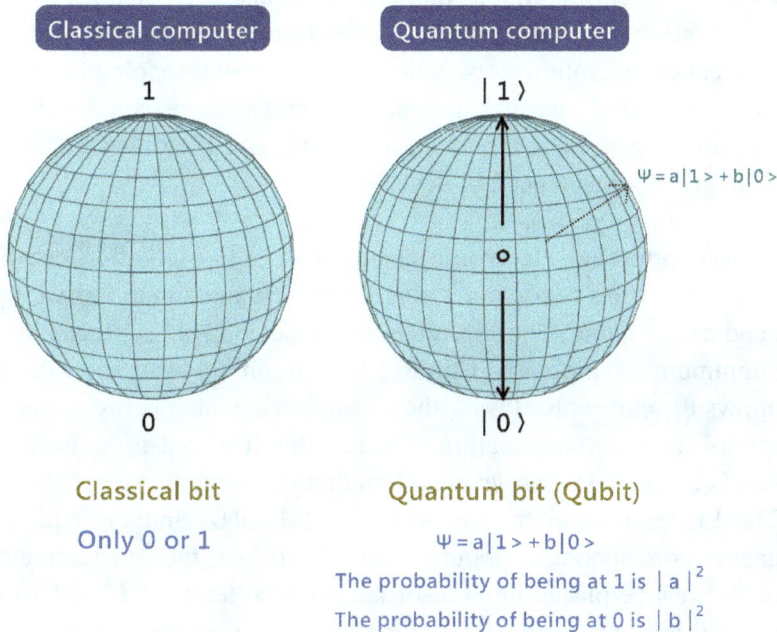

Figure 5.1. Schematic representation of classical bit and qubit. There are infinite sets of possible superposition states for a qubit, each corresponding to a point on the surface of the Bloch sphere.

exponential growth rate is unmatched by classical bits that can only represent 0 and 1. However, it is extremely difficult to increase the number of qubits in a quantum computer from an engineering and technical perspective nowadays. As mentioned previously, it is not easy to turn millions of plates simultaneously, while also requiring each plate to spin at the same rate. It will never be possible to develop the quantum industry without the involvement of high-calibre quantum engineers.

Quantum computing can break through the limitations of classical computing. Quantum computers can perform complex calculations in special ways and at unimaginable speeds, solving some problems beyond the capabilities of supercomputers. However, the potential of quantum computing lies in the uncertainty of quantum states. When interacting with the environment, quantum states will collapse to definite classical states, rendering quantum computing unfeasible. Manipulating qubits without destroying quantum states is a highly delicate and challenging technique. The current qubit fabrication technologies mainly include superconducting electronic circuits, ion traps, cold atoms, nitrogen-vacancy centre, quantum dots, silicon-based qubits, topological qubits, photonic integrated circuits, and nuclear magnetic resonance (NMR). Superconducting circuits, ion traps, cold atoms and photonic technologies are currently more advanced in development.

1. **Superconducting electronic circuits:** An LC oscillation circuit is combined with a superconducting Josephson junction to form a superconducting qubit. The circuit is constructed with superconducting aluminium or niobium. Cooling the circuit to near absolute zero allows the current to flow without resistance, which helps prevent the loss of quantum information. The LC circuit with a Josephson junction is an artificial two-level system that we use to implement a qubit. The LC oscillation frequency is about 1 GHz, and the equivalent energy corresponds to about 48 mK. Therefore, the superconducting qubit must be placed in an environment of at least 10–15 mK to prevent thermal noise from destroying the quantum state. This superconducting artificial atom has many vital advantages. Superconducting quantum circuits are highly compatible with existing integrated circuit systems in design, fabrication, and measurement. Moreover,

traditional electronic components can be used in control systems. Compared with photons and ion traps, superconducting qubits are also easier to manipulate. And, because they are artificial atoms, they are highly scalable, making superconducting qubits one of the most promising technology routes for universal quantum computing. The downside of superconducting qubits is that they must be cooled to near absolute zero using expensive and energy-intensive dilution refrigerators. Otherwise, quantum states will quickly collapse. In recent years, companies such as IBM, Intel, Google, Origin Quantum, SpinQ, QuantumCTek, and Rigetti have invested significant resources in developing quantum technology for superconducting qubits.

2. **Ion traps and cold atoms:** In ion trap and cold atom quantum computing, natural atoms and ions are used as qubits. The atoms and ions are captured using laser cooling techniques in both cases. The main difference is that ions carry a charge, so it is easier to use external electric or magnetic fields to confine ions within a certain region, and to control their movement through the interaction between the charge and electromagnetic fields. Since these atoms (ions) exist in nature and are well understood, the advantage is that coherent quantum states last for a relatively long time in these systems. The logic gate can also be implemented with high fidelity. The disadvantage is that the gate operation speed is slow, laser cooling technology and an ultrahigh vacuum environment are required, and its compatibility with integrated circuits is still to be developed. Furthermore, scalability is another major difficulty in these architectures. At present, leading companies in ion trap quantum computing include Honeywell, quantum start-up IonQ, CIQTEK, Qudoor, Foxconn, and Huayi Quantum. For cold atoms, the companies involved are ColdQuanta and QuEra.

3. **Photonic quantum computing:** Using the quantum properties of photons for quantum computation has significant advantages. The interaction of photons with the environment is very weak so stable quantum states can be maintained for a long time. Therefore, they can be operated under room temperature and an atmospheric environment. However, photonic quantum computing relies on optical elements (beam splitters, phase shifters, etc.) to perform operations, making it relatively difficult to achieve programmability. Practically, photonic

quantum computing is realized by integrating all optical components, including photon sources and detectors, onto a chip using advanced semiconductor technology, which is called a photonic integrated circuit. The other issues of photonic quantum computing are the efficiency of optical quantum memory and integration with traditional integrated circuits. The fact that photons interact with each other very weakly also limits the scalability. Currently, companies developing photonic quantum computing include PsiQuantum, Xanadu, TuringQ, and QBoson.

4. **Silicon-based qubits:** The principle involves using silicon, germanium quantum dots, or ions with spin 1/2 doped in high-purity silicon as qubits. These schemes have good scalability and integration capabilities, and are entirely based on well-established semiconductor manufacturing processes. But, the disadvantage is that the interaction among the quantum dots and the consistency of quantum dots need to be improved. On the other hand, doping impurities in silicon results in a lot of noise, and it is hard to control the position of impurities precisely. The most critical challenge is entangling the qubits. Currently, the number of qubits that can be directly entangled is still low. The main players in silicon-based qubits currently are Intel and Origin Quantum. The best result so far is a 10-qubit result published by the UNSW Australia group in 2022.

5. **Topological qubits:** In recent years, topological quantum computing has been an emerging hot subject that has utilized topological properties in many-body systems to manipulate and store quantum information. The advantages of topological quantum computing include the lack of need for large-scale error correction, high resistance to external perturbations, nearly infinite coherence time, and the almost 100% fidelity of the two-qubit gates. However, currently, the development of topological quantum computing is still stuck in confirming the existence of Majorana zero mode, which is the basic element, not to mention the "braids" required to realize the operation. Microsoft has previously collaborated with the Netherlands in this aspect, but without success. Much progress has been made recently in this direction. For example, in 2016, a team led by Jin-Feng Jia at Shanghai Jiao Tong University observed an indication of Majorana

zero modes in a superconductor/topological insulator heterostructure and, in 2022, a group led by Hong-Jun Gao at the Institute of Physics, Chinese Academy of Sciences, observed a lattice of zero modes in lithium iron arsenide (LiFeAs), taking a big step towards topological quantum computation. Nevertheless, more evidence is still needed and a manipulable system is left to be found. Compared with superconducting qubits, silicon-based qubits, ion traps, and other technologies, topological quantum computing still has a long way to go.

6.   **Nitrogen-vacancy (NV) centres:** If there are nitrogen impurities in synthetic diamonds, substitutional nitrogen atoms may form NV centres with adjacent lattice vacancies. Since the nitrogen of the 5A group has one more electron than the carbon of the 4A group and the lattice vacancy in diamonds also has unpaired electrons, the electrons with spin are tightly trapped in NV centres and can be used as qubits. Due to the stability of the diamond lattice, the spin of NV centres is protected from environmental disturbance. In addition, the characteristic frequency of the NV centre is greatly different from the vibrational frequency of the diamond lattice, so they do not interfere with each other. However, it is difficult to control the location of vacancies in synthetic diamonds. Entangling multiple spin qubits is a great challenge. Quantum Brilliance and CIQTEK are currently actively conducting research in this direction.

7.   **Nuclear magnetic resonance (NMR):** One of the most successful systems for realizing small-scale quantum computing is using nuclear magnetic resonance to manipulate the nuclear spins of molecules. However, unlike other implementations of qubits, NMR quantum computing is performed on an ensemble of systems (large number of nuclear spins) rather than a single quantum state. Therefore, some scientists do not consider NMR implementation to be an example of true quantum computing. Currently, it is possible to operate with seven qubits. In 2001, Isaac L. Chuang's research team used a 7-qubit NMR quantum computer to perform Shor's algorithm to factorize the integer 15 into 3 and 5. Once it reaches more than ten qubits, small-scale quantum advantage calculations can be achieved and commercialized. SpinQ, which provides small desktop-sized quantum

computers specifically for the educational market, is currently developing NMR quantum computers.

The term "Physical Qubits" is usually used to refer to qubits implemented by the previous physical systems. They suffer from decoherence and errors. "Logical Qubits," on the other hand, refer to abstract qubits that are fault-tolerant and can store quantum states stably. The basic unit of a fault-tolerant universal quantum computer is a logical qubit. A logical qubit requires the composition of hundreds to tens of thousands of physical qubits. As long as we can build a quantum computer with 100 logical qubits, it is roughly equivalent to a classical computer with 2^{100} ($\sim 10^{30}$) classical bits. 100 is approximately the number of qubits required for a quantum computer to be more powerful than a supercomputer. If using the 1-nm chip manufacturing process, a classical computer with 10^{30} classical bits is about $10^{10} \times 10^{-9}$ m = 10 m on each side, which can be put in a large room. If there is a quantum computer with 300 logical qubits, it is roughly equivalent to a classical computer with 2^{300} ($\sim 10^{90}$) classical bits. The length of each side would be 10^{21} m, which is far greater than the Earth's radius of 6,371 km. The size of the Earth can only accommodate a classical computer equivalent to a 144-logical-qubit quantum computer (Figure 5.2).

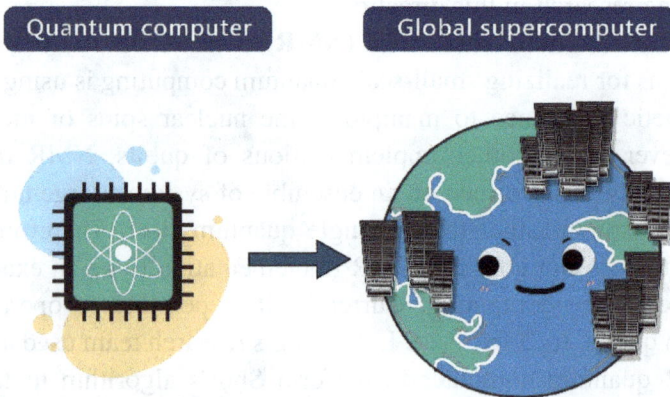

Figure 5.2. A classical computer with computational space roughly equivalent to a 145-logical-qubit quantum computer would be too large to fit on Earth.

5.4 Quantum logic gate and quantum oracle

The core of quantum computing lies in the storage of information in quantum states of matter and the use of quantum gates to process the quantum information. In many quantum algorithms, a component called a quantum oracle appears in the quantum circuit. A quantum oracle is a quantum circuit composed of a sequence of gates and has a specific operational function. Quantum oracles are also known as "black boxes." A "black box" refers to a box whose contents cannot be seen. The internal workings of quantum oracles are unknown; the only thing known is that when you supply inputs, the quantum oracle can provide outputs to achieve specific computing purposes and functions. A key step for the success of quantum algorithms is the efficient construction of a quantum oracle. The following will briefly describe some basics of quantum circuits and quantum gates.

In a quantum circuit, each horizontal line represents the sequence of operations performed on a qubit. As shown in Figure 5.3(a), a quantum circuit consists of three basic processes: the initial state, a quantum gate operation, and the measurement. Two types of quantum gates are required to compose any quantum gate operation. One is single-qubit gates, such as the Hadamard gate, Pauli-X/Y/Z gate, and phase shift gate. The other is two-qubit gates, such as the swap gate and the Controlled-NOT gate. Figure 5.4 shows some commonly used quantum gates and their matrix representation. Quantum gates are reversible unitary transformations, so there is no energy consumption in the computation process until the irreversible "Quantum Collapse" occurs during measurement. Let us take a quick look at a few of the most commonly used quantum gates:

1. **Single-qubit gates: Operations on a single qubit**
   (a) Hadamard gate (H gate)
   The H gate is a fundamental quantum gate that creates superposition states when acting on the $|0\rangle$ and $|1\rangle$ states. When applied to the state $|0\rangle$, it is mathematically represented as $H|0\rangle = (|0\rangle + |1\rangle)/\sqrt{2}$, and when applied to the state $|1\rangle$, it is represented as

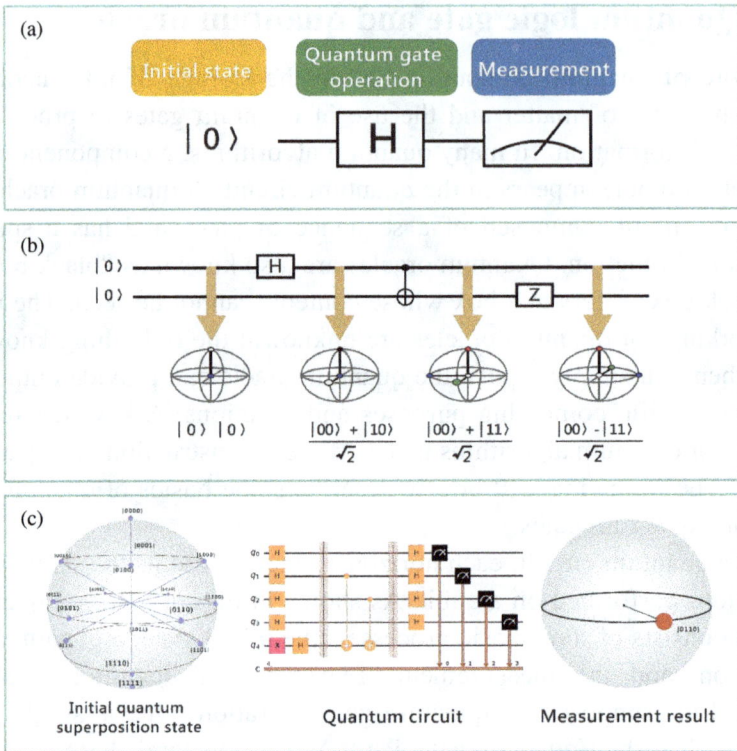

Figure 5.3. (a) Structure of a quantum circuit. (b) A simple example of quantum gate operations on two qubits. Starting from the initial state $|00\rangle$, a Hadamard gate transforms the first $|0\rangle$ into a superposition state of $(|0\rangle + |1\rangle)/\sqrt{2}$. Subsequently, the action of a Controlled-NOT gate (CNOT gate) on the two qubits generates a Bell state of maximum entanglement. The final Pauli-Z gate is equivalent to a rotation around the z-axis of the Bloch sphere. (c) When multiple qubits are involved, there are many horizontal lines and quantum gate operations, and the quantum circuit looks like a staff with musical notes jumping on it.

$H|1\rangle = (|0\rangle - |1\rangle)/\sqrt{2}$. In matrix form, the H gate can be expressed as $H = \frac{1}{\sqrt{2}}\begin{pmatrix} 1 & 1 \\ 1 & -1 \end{pmatrix}$.

(b) Pauli-X gate (quantum NOT gate)

The Pauli-X gate is a quantum analogue to the NOT gate on classical bits, causing the qubit's state to change from $|0\rangle$ to $|1\rangle$ or from $|1\rangle$ to $|0\rangle$. Mathematically, it is represented as $X|0\rangle = |1\rangle$ and $X|1\rangle = |0\rangle$, and its matrix representation is given by $\sigma_x = \begin{pmatrix} 0 & 1 \\ 1 & 0 \end{pmatrix}$.

Operator	Pauli-X(X)	Pauli-Y(Y)	Pauli-Z(Z)	Hadamard(H)	Phase(S,P)
Gate(s)	—X—	—Y—	—Z—	—H—	—S—
Matrix	$\begin{bmatrix} 0 & 1 \\ 1 & 0 \end{bmatrix}$	$\begin{bmatrix} 0 & -i \\ i & 0 \end{bmatrix}$	$\begin{bmatrix} 1 & 0 \\ 0 & -1 \end{bmatrix}$	$\dfrac{1}{\sqrt{2}}\begin{bmatrix} 1 & 1 \\ 1 & -1 \end{bmatrix}$	$\begin{bmatrix} 1 & 0 \\ 0 & i \end{bmatrix}$

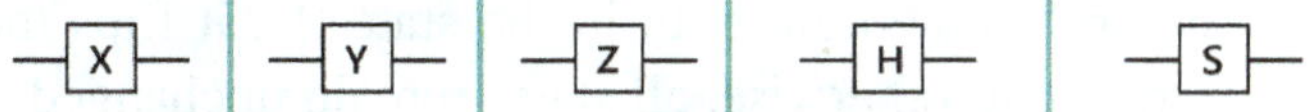

Operator	π/8 (T)	Controlled Not (CNOT,CX)	Controlled Z(CZ)	SWAP	Toffoli (CCNOT, CCX,TOFF)
Gate(s)	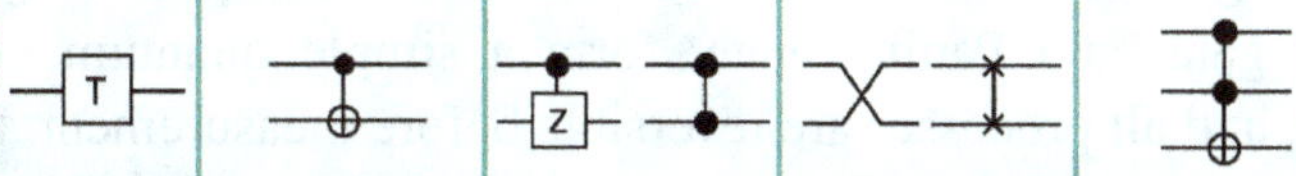 —T—		—Z—	✕	
Matrix	$\begin{bmatrix} 1 & 0 \\ 0 & e^{i\pi/4} \end{bmatrix}$	$\begin{bmatrix} 1&0&0&0 \\ 0&1&0&0 \\ 0&0&0&1 \\ 0&0&1&0 \end{bmatrix}$	$\begin{bmatrix} 1&0&0&0 \\ 0&1&0&0 \\ 0&0&1&0 \\ 0&0&0&-1 \end{bmatrix}$	$\begin{bmatrix} 1&0&0&0 \\ 0&0&1&0 \\ 0&1&0&0 \\ 0&0&0&1 \end{bmatrix}$	$\begin{bmatrix} 1&0&0&0&0&0&0&0 \\ 0&1&0&0&0&0&0&0 \\ 0&0&1&0&0&0&0&0 \\ 0&0&0&1&0&0&0&0 \\ 0&0&0&0&1&0&0&0 \\ 0&0&0&0&0&1&0&0 \\ 0&0&0&0&0&0&0&1 \\ 0&0&0&0&0&0&1&0 \end{bmatrix}$

Figure 5.4. Commonly used quantum gates.

2. Two-qubit gates: Operations involving two qubits

(a) Controlled-NOT gate (CNOT gate)

Among the two involved qubits, one is the control qubit and the other is the target qubit. When the control qubit is in the state $|1\rangle$, it performs a Pauli-X gate on the target qubit, changing the state of the target qubit from $|0\rangle$ to $|1\rangle$ or vice versa. However, when the control qubit is in the state $|0\rangle$, the target qubit remains unchanged. Therefore, the action of a CNOT gate is mathematically represented as $\text{CNOT}|00\rangle=|00\rangle$, $\text{CNOT}|01\rangle=|01\rangle$, $\text{CNOT}|10\rangle=|11\rangle$, and $\text{CNOT}|11\rangle=|10\rangle$. The matrix form is

$$\text{CNOT} = \begin{pmatrix} 1&0&0&0 \\ 0&1&0&0 \\ 0&0&0&1 \\ 0&0&1&0 \end{pmatrix}.$$

In Chapter 4, we introduced a key quantum gate in reversible computing, the Controlled-Controlled-NOT (CCNOT) gate, which has three inputs and three outputs. If the first two qubits (control qubits) are both in the state $|1\rangle$, it flips the state of the third qubit; otherwise, all states remain unchanged.

Figure 5.3(b) shows two qubits with initial states of $|0\rangle$, which are subjected to different quantum gate operations. First, the upper qubit is transformed into a superposition state after being acted on by an H gate, and then forms an entangled state with the lower qubit through the action of a CNOT gate. Next, the entangled state of the upper and lower qubits is subjected to an action of the Pauli-Z gate on the lower qubit, causing the entangled state to rotate along the Z axis. Three quantum gates, H gate, CNOT gate, and Pauli-Z gate, form a simple quantum circuit called Oracle, and all processes are reversible before measurement. When multiple qubits are involved, there are many horizontal lines and quantum gate operations, and the quantum circuit looks like a staff with musical notes jumping on it. Therefore, some people compare writing a quantum circuit programme to composing, and quantum computation to a symphony, with qubits playing their melody in harmony with each other.

5.4.1 *Matrix multiplication*

Matrix multiplication is the operation of multiplying two matrices to obtain a matrix product. Matrix multiplication is a basic mathematical operation in Hilbert space. The function of the quantum gate on a quantum state can be clearly expressed by matrix multiplication. Readers interested in learning more about mathematics can read relevant mathematical books to enhance their basic knowledge. We will only use the example shown in Figure 5.5 to illustrate the operation of matrix multiplication. In classical Newtonian mechanics, vectors in three-dimensional space have three vector components, but as introduced in Chapter 3, n qubits can span a vector space of 2^n, meaning there are 2^n unit vectors. This high-dimensional Hilbert space provides a major advantage to quantum computing over classical computing. In Figure 5.5, five Chinese characters are shown, 金(gold), 木(wood), 水(water), 火(fire), and 土(soil), representing the five

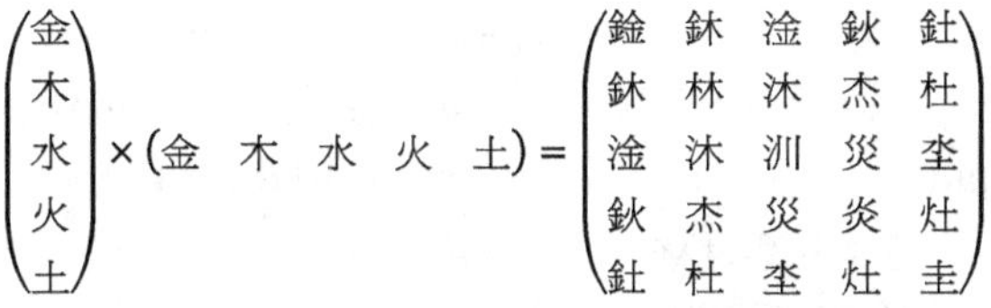

$$\begin{pmatrix} 金 \\ 木 \\ 水 \\ 火 \\ 土 \end{pmatrix} \times \begin{pmatrix} 金 & 木 & 水 & 火 & 土 \end{pmatrix} = \begin{pmatrix} 鎽 & 鉢 & 淦 & 鈇 & 釷 \\ 鉢 & 林 & 沐 & 杰 & 杜 \\ 淦 & 沐 & 泖 & 災 & 坔 \\ 鈇 & 杰 & 災 & 炎 & 灶 \\ 釷 & 杜 & 坔 & 灶 & 圭 \end{pmatrix}$$

Figure 5.5. An example of matrix multiplication. The five Chinese characters, 金(gold), 木(wood), 水(water), 火(fire), and 土(soil), represent the five vector components of two five-dimensional vectors. The matrix multiplication of a column vector (5 × 1 matrix) and a row vector (1 × 5 matrix) results in a 5 × 5 matrix with 25 new Chinese characters.

vector components of two five-dimensional vectors. These characters are actually from the "five phases" of the Chinese philosophy *Wuxing*. According to the definition of matrix multiplication, the corresponding matrix elements are obtained, as shown in the results in Figure 5.5. These 25 elements are actually Chinese characters with different meanings.

Qubits can exist in an entangled state, thus having strong correlations. As a result, operations performed on a qubit in a quantum computer will simultaneously affect multiple other qubits entangled with it. This is different from parallel computing on classical computers. Classical bits are essentially operated independently, and each bit flip does not affect other bits. In parallel computing, even if multiple processors are processed in parallel, each processor still only does one thing at a time; the multiple processors simply work together by dividing the work. In opposition, because qubits can be in superposition and entangled states, if there is a 100-qubit quantum computer, each operation can potentially affect 2^{100} different states. 2^{100} combinations of states are already more than the number of known atoms on the Earth. Why is quantum simulation of physical systems feasible for a quantum computer? Because the key advantage of qubits is that through quantum superposition and entanglement, they can expand the dimension of computational space at an exponential rate and process a large number of states simultaneously. This "quantum parallelism" is more like the real parallel computing.

If a person walks 1 m at a time, how far can he go after walking 30 times? We all know the answer is 30 m. This is the rate at which classical bits expand the computational space dimension. If there is a quantum computer with 30 qubits, the dimension of the computational space would be 2^{30}. If using the same analogy, the person has already circled the Earth 26 times. This "exponential change" is the true power of quantum computing. Each additional qubit doubles the state that a quantum computer can store. For a quantum computer with 72 qubits, the number of established quantum states is equal to the encoding of 2^{72} classical bits, which requires about 26 Summit-level supercomputers to realize. Since quantum computers are capable of processing vast amounts of information, they can solve the simulation problem of complex quantum systems that are incredibly difficult for classical computers, which was why Feynman and Deutsch initially proposed the idea of quantum computers.

5.5 Cryoelectronics

Qubits are extremely sensitive to their environment. The qubits in a quantum computer must be in the quantum state of superposition and entanglement to offer various performances beyond those of classical bits, and temperature is one of the important factors that can destroy the quantum state. In principle, qubits must be placed in an isolated, ultra-low-temperature environment (~mK or ~−273°C) that is as free from external interference as possible. The working conditions of existing semiconductor components are mostly around room temperature, so the design parameters of electronic components are very different from those of quantum components. Therefore, cryoelectronics has become an essential field for the development of quantum technology in the future. For semiconductors, the generation of electron–hole pairs becomes more active as the temperature increases, and the actual operating temperature varies with the type of components and materials. But, at extremely low temperatures, many of the currently used semiconductor materials become insulators, and therefore the design parameters of traditional electronic components in the mK (~−273°C) range must be re-examined. Cryoelectronics focuses on the study of electrical properties of materials and semiconductor devices and their applications in the temperature range from 77 K (−195.79°C) to absolute zero (−273°C). Superconductor electronics is also included in the scope of cryoelectronics. The field of cryoelectronics emerged from the combination of cryogenic engineering and power electronics. Cryoelectronics considers the changes in various electronic components after a significant reduction of thermal noise. The current research directions include quantum parametric amplifiers and various cryogenic instruments and devices, the properties of materials and electronic components under cryogenic conditions, and test instruments required for superconductor electronics.

While qubits must be operated at extremely low temperatures (~−273°C), the control system and readout signals still need to be transmitted back to room temperature. Therefore, many novel cryoelectronics designs are required in the conversion process from room temperature to extremely low temperature. This is a patent-blank area that was not covered by traditional semiconductor technology. It is not only a key technology for quantum computing but also a battleground of start-ups for quantum computing patents. Horse Ridge, a cryogenic control chip designed by Intel, effectively separates electronic components at 1.1 K (~−272°C) from qubits at mK (~−273°C), simplifying the quantum chip manufacturing process, and is also a key technology for the scalable fault-tolerant quantum computers.

Tech giants such as IBM and Google have invested in superconducting electronic circuits, which are leading candidates for realizing large-scale quantum computing. The rest of this section will mainly focus on superconducting quantum computers. One can control superconducting qubits at low temperatures by applying microwave pulses, so understanding the principles of the interaction between light and matter is essential to designing circuits. The most natural qubits are atoms and photons, and studying the interaction between light and matter involves studying how photons and atoms interact; this field is called Quantum Electrodynamics (QED), or specifically Cavity Quantum Electrodynamics (CQED) if photons are trapped in a box with mirror walls. The interaction between superconducting qubits and microwaves imitates the interaction mechanism between atoms and photons in nature. The physical and mathematical problems of superconducting quantum circuits are basically the same as those of CQED, so the problems in this area of superconducting quantum circuits are also known as circuit QED (cQED). Problems addressed within circuit QED are often hybrids of classical circuits and quantum systems because other electronic components must be integrated. More quantum engineers are needed to improve the susceptible and fragile quantum state of qubits.

The architecture of a quantum computer is similar to that of a classical computer. It requires the combination of many different levels of software and hardware design to create a commercial quantum computer. What is more complicated than classical computers is that quantum states in quantum computers must be carefully maintained. Therefore, the working environment for quantum computers is completely different from the current semiconductor technology and new technologies need to be developed. In addition, integration with classical computers is necessary for practical uses (Figure 6.5 (a)). The current superconducting quantum computing system can be divided into at least the following three regions (Figure 5.6 (b)):

Extremely low-temperature region — The quantum layer: The extremely low-temperature region with a temperature near 10 mK (~−273°C) is where the superconducting qubits are located and quantum logic gates are operated. The critical technology of quantum computers is primarily in this region. At present, the main competition between the world's foremost companies and developed countries is to master quantum technology in this extremely low-temperature region. For superconducting qubits to operate near absolute zero, an excellent and efficient cooling system is essential. Currently, only few companies can commercially produce this kind of cooling system; Oxford Instruments in the UK and Bluefors in Finland are the top sellers. Recently, both China and IBM started independently developing their own large-scale cooling systems to prepare for the cryogenic environment required for more qubits in the future. It is estimated that at least 20% of the component market for quantum computers belongs to the core technology of this extremely low-temperature region.

Low-temperature region — The control and measurement layer: The main function of this region is to control the qubits, and it is the signal conversion layer between the quantum layer and the control processor layer. The temperature in this region varies from 10 mK (~−273°C) to

(a)

Classical input

Quantum input

Classical output

Quantum
measurement output

Digital program

Quantum (gate) circuit

(b)

Applications

Algorithm and software

Classical computer

high speed digital signals

Near room-temperature region
(4K ~ room-temperature)

Room-temperature electronics
Control circuits

Quantum
–Classical interface

Low temperature region(mK ~ 4K)

analog signals

Cryoelectronics / Control circuits

Quantum plane
extremely low temperature region(~mK)

Physical and logical qubits

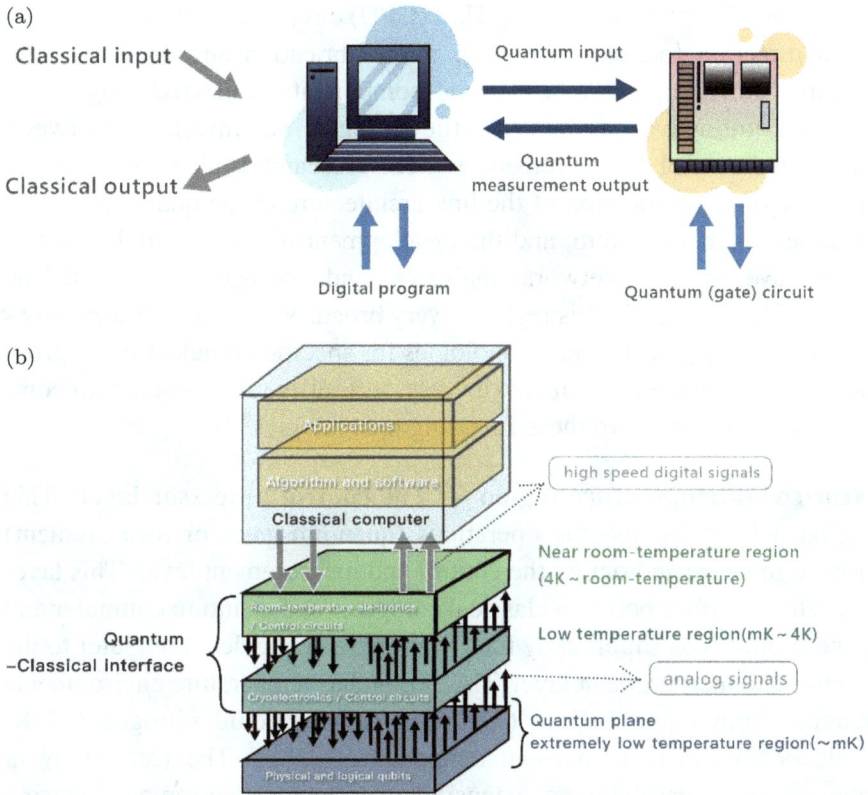

Figure 5.6. (a) Integrated system of quantum computer and classical computer. (b) Layering of quantum–classical integrated system.

around 4 K (−269°C), which is liquid helium temperature. The technology in this region, especially the integration between qubits, control systems, and measurement systems in a low-temperature environment, is a vital area of cryoelectronics. As qubits are extremely sensitive, an engineering problem that needs to be solved urgently is bringing the control system as close as possible to the qubits for effective operation without affecting the temperature of their working environment. This is an important development direction for future scalable quantum computers. In addition, various fields of electronics need to be extended to the case of the low-temperature range to reconsider design parameters and material properties. This aspect includes low-temperature CMOS components and various ADC, DAC,

FPGA, and Graphics Processing Unit (GPU) circuit designs. The development direction focuses on precise microfabrication and creating more effective low-temperature electronic components while reducing costs. Besides intensive studies of the integration interface between qubits and control (measurement) systems, research is also conducted on the design and production of the link architecture of the quantum control and measurement system, and the development of filters, amplifiers and microwave sources, network analyzers, and spectrum analyzers. The scope of technology in this region is very broad. Many start-up companies are developing patents and technologies for specific cryoelectronics problems. It is estimated that approximately 50% of the future quantum computer market belongs to these fledgling technologies.

Near-room-temperature region — The control processor layer: This region defines the specific operations (quantum gates or measurement) needed to be carried out by the control and measurement layer. This layer plays the interface between classical computer and quantum computer and issues commands supplied by the programme in classical computer to the control and measurement layer. This region has temperature environments ranging from liquid helium (4 K, −269°C) to liquid nitrogen (77 K, −195.79°C) and then to room temperature (25°C). The technology in this region is basically an extension of existing electronics, focusing on control circuits and system integration. Since liquid helium and liquid nitrogen temperatures are relatively easy to reach, there have already been some special market demands for electronic applications in this temperature range in the past. However, efficiently transferring large amounts of information at 4 K back to the classical computer at room temperature for processing requires further development of existing electronic component technology. It is estimated that these extensions account for about 30% of the future quantum computer hardware market.

At present, the development of quantum computers is still in the rather early stages. Still, the integrated architecture of quantum computer and classical computer has been considered to be the optimal computing system when it matures in the future. This type of integrated system can be functionally divided into three layers: the quantum computer layer, the quantum–classical interface layer, and the classical computer layer.

The quantum computer layer corresponds to the extremely low-temperature region we just mentioned. The temperature of this layer must be kept at 10 mK to prevent quick decoherence of quantum states. This layer involves critical technology for manufacturing qubits, which is the core of the system. The quantum–classical interface layer includes the low-temperature region and a part of the near-room-temperature region. This layer identifies the sequence of operations and measurements needed to implement the quantum algorithm. The conversion of digital signals of the control processor into analogue signals for the quantum computer layer is also performed in this layer. The quantum–classical interface also converts the analogue measurement outputs of the quantum computing layer into binary and transfers them back to the classical computer layer. This transmission process is crucial for the execution of iterative operations and quantum error corrections. It limits the operation speed of the quantum computer and the whole integrated system. Improving the transmission process and interacting with the quantum computer layer without disturbing the quantum state are the key development directions of the quantum–classical interface layer. The technology at this layer mainly involves the application of cryoelectronics, which is a highly competitive area for quantum start-ups. The classical computer layer is responsible for reading the results from the quantum–classical interface layer, interacting with large storage arrays and networks, and analyzing and relaying commands of the user-friendly interface. The technology used here is similar to current semiconductor electronics. Many start-ups and research teams are working on creating better and more user-friendly quantum–classical integrated systems. For example, in 2022, Nvidia announced the development of an integrated architecture of classical computer and quantum computer that focuses on establishing a fast, low-latency bus between the GPU and the Quantum Processing Unit (QPU), accelerating classical computing tasks such as circuit optimization, qubit calibration, and error correction through the GPU.

5.6 Quantum electronic design automation

Electronic design automation (EDA), known as the "mother of chips," is an indispensable part of IC design. International companies such as

Synopsys, Cadence, and Mentor have long monopolized the global EDA market, which is one of the reasons why the development of China's semiconductor industry has long been stuck in a bottleneck. The core branch of EDA is Technology Computer Aided Design (TCAD) tools, which are programmes that simulate semiconductor fabrication processes and device properties. In terms of quantum computers, because the physical characteristics and temperature environments are vastly different from classical computers, existing EDA software is incapable of designing quantum chips. At present, quantum EDA (QEDA) is still in the early stages of research and development. Sandia National Laboratories in the United States, IBM and IQM in Finland, and Delft University of Technology in the Netherlands are all actively working on quantum EDA. Canadian company Nanoacademic Technologies has released a commercial quantum TCAD tool, QTCAD. China's Origin Quantum developed the quantum chip design software "Origin Unit" in 2022, hoping to overtake the competitors in the early development of QEDA and avoid a situation where EDA technology is controlled by others like in the past.

5.7 Different types of quantum computers

Quantum computers can be divided into four types:

1. **Universal quantum computer:** A universal quantum computer is a quantum computer that can execute arbitrary quantum algorithms and is composed of many logical qubits. The quantum circuit is widely used in quantum computing, which refers to the performance of a series of operations on qubits to achieve quantum computation. These operations include initialization of qubits, unitary transformations of quantum states, and reading information of qubits. To perform quantum computing to solve particular problems, we need to build a specific quantum circuit using various quantum gates and execute it on a universal quantum computer. For example, to factor large numbers, we construct a quantum circuit for Shor's algorithm and execute it. To perform a database search, we execute a quantum circuit of Grover's algorithm. At present, there are not many quantum algorithms, and they can only speed up specific issues. The essential thing in quantum

computing is to recognize which problems are suitable for quantum computers and design quantum algorithms that perform better than classical algorithms. It is indeed a great advantage to use a universal quantum computer, which harnesses quantum properties of superposition and entanglement, to execute quantum algorithms to solve specific problems. However, the current challenge is to build a fault-tolerant quantum computer. According to DiVincenzo's Criteria proposed in 2000, a quantum computer must meet the following five conditions: scalability, the ability to initialize the state of the qubits to a simple fiducial state, long relevant decoherence times, the ability to implement a set of universal quantum gates, and the capability of qubit-specific measurement. IBM has also used a metric called Quantum Volume (QV) to evaluate the performance of quantum computers. QV is related to the number of qubits, the connectivity of qubits, the length of coherence time, the fidelity of quantum gate operations, and the measurement error. The larger the quantum volume, the stronger the quantum computer, and the more likely it can solve complex real-world problems such as chemical problems, drug discovery, and asset allocation in finance. Both IBM and Honeywell claim that around 2025 they will launch a quantum computer with a sufficiently high quantum volume to solve practical social and natural problems. In 2021, IBM further proposed an indicator related to the speed at which a quantum processor executes quantum circuits, called Circuit Layer Operations Per Second (CLOPS). IBM will also announce the 100 × 100 Challenge with Heron in 2024. Heron will provide CNOT gate entangled 100 qubits and depth-100 gate operations in a reasonable runtime with error free and this with complexity and runtime will beyond the capabilities of the best classical computers today.

2. **Special-purpose quantum computer:** A special-purpose quantum computer can only execute specific quantum algorithms. If one needs to perform calculations beyond the originally designed function, the hardware or equipment must be changed, which is entirely different from a universal quantum computer, where the executed programme can be changed at will. The quantum computer that D-Wave uses to perform quantum annealing to find global optimal solutions and China's "Jiuzhang" used to perform Gaussian Boson Sampling belong

to this type of specific function quantum computer. Special-purpose quantum computers mainly fall into two categories: quantum anneal-ers and quantum analogue simulators. A quantum computer that utilizes the quantum tunnelling effect of quantum components to implement quantum annealing algorithms to solve combinatorial optimization problems is called a quantum annealer. QBoson in Beijing launched the 100-qubit Coherent Ising Machine (CIM) "Tiangong Quantum Brain" in 2023. Compared with the classical optimization algorithm, CIM's average acceleration ratio is 100 times greater and the average solution result is 120% better. Ping An Bank of China has used this quantum CIM to perform feature screening of a German credit dataset in <1 millisecond. A quantum analogue simu-lator is a controllable quantum system that is artificially designed to simulate the natural quantum system in nature. Recently, the Silicon Quantum Computing (SQC) team in Australia created an atomic-scale quantum integrated circuit to simulate the quantum state of an organic polyacetylene molecule, demonstrating the effectiveness of artificial quantum simulators. In the future, quantum analogue simulators could be used to discover new materials that have never existed before to develop new drugs, battery materials, and catalysts.

3. **Quantum-inspired computer:** A quantum-inspired computer is a computer that uses traditional electronic components to simulate quantum tunnelling behaviour and is specifically designed to handle combinatorial optimization problems. Semiconductor technology is used to create special-function chips that simulate quantum algo-rithms, which can work at room temperature. Companies such as Fujitsu, Toshiba, and Hitachi have made operable chips and machines.

4. **Educational quantum computer:** NMR-based and NV centre-based 2-qubit quantum computers have been commercialized for educational purposes. Their main advantage is that they can operate at room tem-perature. Their disadvantage is that they only have two qubits, and thus cannot perform any calculations of practical significance, but they can demonstrate basic operations in quantum computing for educational purposes. SpinQ has already provided several versions of NMR-based educational quantum computers on market at reasonable prices.

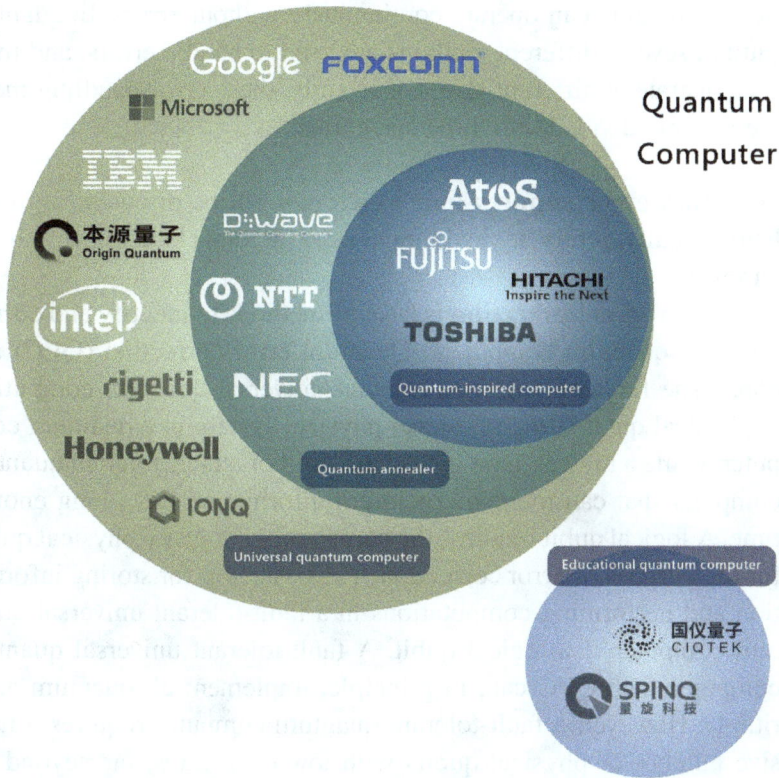

Figure 5.7. Different types of quantum computers and vendors.

IQM recently announced an educational and research quantum computer with 5 superconducting qubits for less than $1M Euro.

Figure 5.7 summarizes various types of quantum computers along with their vendors.

5.8 Quantum error correction and quantum fault tolerance

At present, most of the physical qubits can only maintain their quantum states for a very short time, much shorter than the time that transistors in classical computers can operate continuously without errors. In quantum computing, several different methods are used to handle errors, and these methods operate at different levels. Currently, these error-handling methods are roughly divided into three categories:

1. **Quantum error suppression**, which will not be discussed in detail here, usually refers to handling errors at the closest level to the hardware.
2. **Quantum error correction** is a set of methods to identify and correct errors in quantum computers. Quantum error correction (QEC) is a core issue for realizing fault-tolerant universal quantum computing. A physical qubit is the two-level physical system in a quantum computer, while a logical qubit is the basic unit of a fault-tolerant quantum computer that can maintain quantum information for a long enough time. A logical qubit requires the composition of many physical qubits for the purpose of error correction. The basic unit for storing information and performing computations in a fault-tolerant universal quantum computer is a logical qubit. A fault-tolerant universal quantum computer with QEC can, in principle, implement all quantum algorithms. However, a fault-tolerant quantum computer requires a massive number of physical qubits with low error rates, far beyond the current technology. In fact, with the current technology, even one logical qubit has not yet been achieved.

Decoherence of quantum states leads to bit errors and phase errors. Bit errors cause changes in the squared modulus of coefficients of $|0\rangle$ and $|1\rangle$, while phase errors cause changes in the phase between $|0\rangle$ and $|1\rangle$. For example, for a superposition state $a|0\rangle+b|1\rangle$, a bit error makes the state change to $c|0\rangle+d|1\rangle$, and phase error can result in a state of $a|0\rangle-b|1\rangle$. Bit errors are also present in classical computers, but phase errors are unique to quantum computers. In 1997, Alexei Kitaev proposed topological codes, also known as surface codes. Theoretically, surface codes are currently the best choice for quantum error correction. However, when the surface code is practically applied, it may suffer from interference from nearby qubits. Therefore, IBM developed the heavy hexagon code, which combines the advantages of the surface code and the Bacon–Shor code. The disadvantage is that the connectivity of qubits is reduced in this layout. Nevertheless, the heavy hexagon code is currently the best compromise between theory and practical application. Although it is more complicated than the surface code in topology, it has significant benefits for error correction. Most importantly, compared to square geometry, the heavy hexagon code greatly reduces the probability of frequency collisions, which are unwanted interactions with neighbouring qubits. The heavy hexagon code significantly improved the error rates of IBM's device during practical application. Moreover, the heavy hexagon code is scalable, making it a breakthrough in building large-scale universal quantum computers. A fault-tolerant universal quantum computer is a long-term development goal that will take some time to achieve, but recently IBM developed the Eagle quantum chip with 127 qubits and the Osprey quantum chip with 433 qubits. The number of classical bits required to simulate Eagle is roughly equal to the total number of atoms in all the people on Earth. In other words, the number of configurations that the Eagle chip can represent is already beyond the capability of supercomputers.

3. **Quantum error mitigation:** The term "NISQ device" was coined by quantum information scientist John Preskill, referring to quantum computers with 50 to several hundred of qubits, which still suffer from quantum decoherence and noise in quantum gates. However, a current quantum computer with more than tens of qubits is already more powerful than supercomputers at performing specific tasks. Therefore, conducting research on NISQ computing is still of great significance. For NISQ devices, in addition to developing corresponding quantum algorithms, it is also necessary to address the issue of errors in computation. Due to the limited number of qubits, a quantum error correction scheme for forming logical qubits is obviously not applicable in the NISQ era. Therefore, recent research has focused on using quantum error mitigation techniques in quantum computations. Quantum error mitigation methods improve the computation results by post-processing measured outputs of ensembles of circuits. Currently, commonly adopted error mitigation strategies are (i) probability successful trial, (ii) error extrapolation, and (iii) modulus scaling. Restricted by the number of qubits that current technologies can

provide, NISQ computing is the only possible application of quantum computers in the near future. With the help of quantum error mitigation, we can perform quantum computations that do not require too many quantum gate operations. For example, variational quantum algorithms, a hybrid quantum–classical algorithm that uses a classical computer to optimize a parameterized quantum circuit, can be implemented within these constraints and used to solve some important problems in quantum chemistry and materials science that are difficult for classical computers. However, since variational quantum algorithms involve large-scale parameter optimization and their performance strongly depends on the initially selected trial quantum circuit (ansatz), it is still only a heuristic method. That is to say, its quantum advantage over classical algorithms cannot be rigorously proven.

5.9 Concluding remarks

Quantum technology has emerged in recent years, leading to earth-shaking changes in the world. IBM's quantum computing roadmap indicates that around 2030, fault-tolerant universal quantum computers will appear, which is the arrival of the era of quantum computing, or Y2Q. Recently, quantum computing as a service (QCaas) provided by NISQ devices and quantum annealers has begun to show quantum advantages in many application areas. Quantum computers and computing are now reaching a near-commercial state, but many difficult challenges remain. Quantum scientists all over the world are working on improving the consistency in qubit quality, the fidelity of quantum gate operations, and the scalability of quantum systems. These are all crucial for achieving large-scale commercial applications. Various physical implementation schemes of quantum computer are still in a state of competition. Although superconducting qubits and ion trap techniques are currently in the leading position, the ultimate result and final winner are yet to be proven. On the other hand, cryogenic control technologies are vital in preventing the influence of thermal noise in the control and transmission of quantum information. Therefore, effective and inexpensive cryogenic technologies need to be developed. Quantum computing also requires specific algorithms. The existing algorithms are insufficient to support large market demand, and new and more effective quantum algorithms are still waiting to be developed. In addition, the technology of quantum computers is distinct from that of classical computers. It remains to be seen whether the development of quantum computers can follow a Moore's Law-like trajectory as classical computers have. Quantum technology has already transitioned from the research to the engineering stage, which requires many engineers to participate in the development work. It is encouraging to see some commercial business already initiated into the field of educational quantum computers, e.g., IQM and SpinQ. IBM has also delivered several quantum computers for application purposes worldwide. In the future, various industries related to quantum technology will also gradually mature and improve due to the investment of human and material resources from multiple countries.

It should be emphasized again that quantum computing outperforms classical computing only on a few specific problems. The main task of quantum computers is not simple arithmetic operations or text input, but instead dealing with NP-hard problems that are difficult for classical computers to solve in a reasonable amount of time. By mapping specific mathematical models to quantum systems or applying quantum logic gates, quantum computers can harness quantum superposition and entanglement to efficiently solve complex problems such as large-number factorization, quantum chemistry, and quantum artificial intelligence. Practical quantum computers require a large number of low-error qubits; therefore, a further breakthrough in quantum technology is necessary. Quantum computers are not intended to replace classical computers, but complement them. The ideal scenario is to have quantum computers and classical computers work together and each handle tasks where they have the advantage. Generally, classical computers perform better in tasks such as downloading web pages and graphics processing for computer games, but quantum computers have an advantage in areas such as machine learning for image and speech recognition and processing of big data. Quantum computing can perform real-time analysis of complex data obtained from information flow. Then, these valuable data can be converted into user-friendly results through classical computers, providing users with a basis for real-time decision-making. Future competitive advantages will likely go to companies integrating quantum and classical computing. Guo Guoping, a professor at the University of Science and Technology of China, stated, "Building high walls, accumulating grain, and becoming king slowly" is the strategy to promote the quantum computing industry. "Building high walls" means building a solid wall of quantum computing technology; "accumulating grain" means connecting resources of engineering and applications; and "becoming king slowly" means not rushing to go public and realize profits. China's current scientific and technological environment is mature and human resources are sufficient. In the invisible war of the "Second Quantum Revolution," "going forward together and fighting steadily" is the mentality that all practitioners working hard in the field of quantum technology must have.

The quantum computer is an application tool that, once mature, will disrupt almost all industries and make great contributions to finance, military, intelligence activities, environmental engineering, deep space exploration, drug design and discovery, aerospace engineering, nuclear fusion, polymer design, artificial intelligence, big data, AI chatbot, and other fields. In the future, quantum computers will not only solve the most complex problems and mysteries in human life but will also serve as the brain of robots. Now is the best time for industry, academia, and governments to think about the positioning of quantum computers and their impact on the future. Even though the number of physical qubits on a IBM Condor quantum computer has exceeded 1121, the quantum circuit that can be executed today is far from deep enough to implement quantum algorithms due to the short coherence time. Therefore, although a few quantum algorithms have been proven to have advantages in specific problems, there is still a way to go before practical applications can be achieved. The error rate of qubits in the current quantum computer is still high. Although error correction methods and fault-tolerant quantum computation can be used to remedy it, the cost of auxiliary qubits is enormous. Using as an example the fault-tolerant quantum computation based on surface codes in superconducting quantum systems, even if the error rate of a single quantum gate operation is as low as 0.1%, at least 1,000 to 10,000 physical qubits are required to encode a logical qubit, resulting in the difficulty of building a fault-tolerant quantum computer and implementing quantum algorithms. The good news is that before the advent of a fault-tolerant universal quantum computer, applications of NISQ have gradually emerged and shown advantages. In addition, the development of hardware is also advancing by leaps and bounds. IBM has announced that its "Kookaburra" quantum computer with over 4,000 physical qubits will be available by 2025, along with quantum error correction and an integrated classical and quantum computing system. The double exponential growth rate of Neven's Law for computational power of quantum computers is much faster than Moore's Law for classical computers. Many existing technology industries will undergo tremendous changes after the "Second Quantum Revolution," and we must be prepared to enter a brand new quantum era.

© 2024 World Scientific Publishing Company
https://doi.org/10.1142/9789811287015_0006

Chapter 6

Quantum Algorithm

If you cut circumference of a circle very finely, you will only lose very little. If you cut it again and again until it cannot be cut, then it will be integrated with the circumference and nothing will be lost.

— Liu Hui (225–295)

A classical computation is like a solo voice — one line of pure tones succeeding each other. A quantum computation is like a symphony — many lines of tones interfering with one another.

— Seth Lloyd (1960–)

6.1 What is an algorithm?

Many commonly used mobile apps, such as YouTube, Instagram, and Facebook, have their own unique algorithms and special functions. Users often find some software more user-friendly; the main difference lies in the quality of their algorithms. But, what exactly is an algorithm? An algorithm is a step-by-step instruction for solving a specific problem. The Euclidean algorithm commonly used in middle school is a prototypical example, and a computer is a universal machine that can execute any algorithm. The word "Algorithm" comes from "Algoritmi," the Latin translation of Al-Khwarizmi, the name of a great Persian mathematician of the ninth century. Al-Khwarizmi used standardized rules of "substitution" and "elimination" to solve algebraic equations. The rule that defines a series of operations is now referred to as a "programme" or "algorithm" in modern times. In ancient China, algorithms were called "術" (translated as art) and first appeared in the book *On the Gnomon and the Circular Paths of Heaven* 《周髀算經》 and *Nine Chapters of Arithmetic* 《九章算術》. In particular, *Nine Chapters of Arithmetic* provides algorithms for calculating elementary arithmetic, the greatest common divisor, the least common multiple, square roots, cube roots, and solving linear equations. Another example is Liu Hui's π algorithm in the Three Kingdoms period, which is a method of calculating π to arbitrary accuracy using an iterative procedure. Liu Hui observed that as the number of sides of a polygon inscribed in a circle increases, the area of the polygon approaches that of the circle. Using his algorithm, Liu Hui first divided the circle into a 192-sided regular polygon and estimated the π value to be 157/50 = 3.14. He then further calculated the area of a regular polygon with 3,072 sides, obtaining a value of π equal to 3,927/1,250 = 3.1416, which became known as Hui's Ratio. Liu Hui knew that dividing a circle into more regular polygons could more accurately approximate the value of π, for which he had already used the concept of limit in modern mathematics. However, he did not have the right tools to realize this method as many times as he could. Even with algorithms, one cannot complete specific computing tasks without proper implementation tools. The advent of computers has provided a platform to truly demonstrate the potential of various algorithms.

Ada Lovelace Byron was the first person to recognize the potential of computers, arguing that computers can not only count like fingers or an abacus but also perform complex calculations. She was also the earliest programmer and once published an algorithm for solving Bernoulli numbers in mathematics. Algorithms contain instructions for how computers process information. Algorithms are used to perform a specified task, which can be as simple as calculating grades and printing papers, or as complex as solving integral equations and even assisting managers in making decisions. Starting by reading data from input devices, processors process data according to a sequence of instructions of an algorithm and the results are written to an output device. Without algorithms, computers cannot process anything, regardless of how powerful they are. In short, an algorithm is composed of a finite sequence of instructions, used to solve a specific problem. For example, if the problem at hand is how to cook frozen dumplings, the algorithm would be a recipe for cooking frozen dumplings, which requires the following steps:

A. Prepare input information
Step 1. Take the frozen dumplings out of the refrigerator.

B. Prepare tools for implementation
Step 2. Prepare a pot for cooking.

C. Algorithm
Step 3. Add enough water into the pot and turn the stove on high heat until the water boils.
Step 4. Put the frozen dumplings into the pot.
Step 5. After the water boils again, add a little cold water. Repeat this step three times.

D. Output results
Step 6. Place the cooked dumplings on a plate.

As shown in Figure 6.1, step 1 is like reading data from an input source, and step 6 is like outputting the results to the output device. Step 2 involves choosing the hardware that implements the algorithm. Due to differences in heating efficiency and capacity, using a pan or a wok will result in a different taste for the dumplings. Therefore, it is important to have the right tools for the job. Steps 3–5 are the algorithm for cooking dumplings, with step 5 being the important iterative step for making delicious dumplings. The purpose of repeating the addition of cold water three times is to lower the temperature of dumplings through cold water. There are two purposes for cooling the water in step 5. One is to heat evenly to avoid overcooking the outer dumpling skin while the inner filling remains undercooked. Another reason is to prevent the dumpling skin from becoming gelatinized at high temperatures. Slightly reducing the temperature can make the dumpling skin more elastic and increase the taste. As long as the algorithm is clear and effective, anyone can cook delicious dumplings according to the algorithm. Most people know how to cook dumplings and know that they need to add cold water three times, but they do

Recipe/Algorithm for cooking frozen dumplings

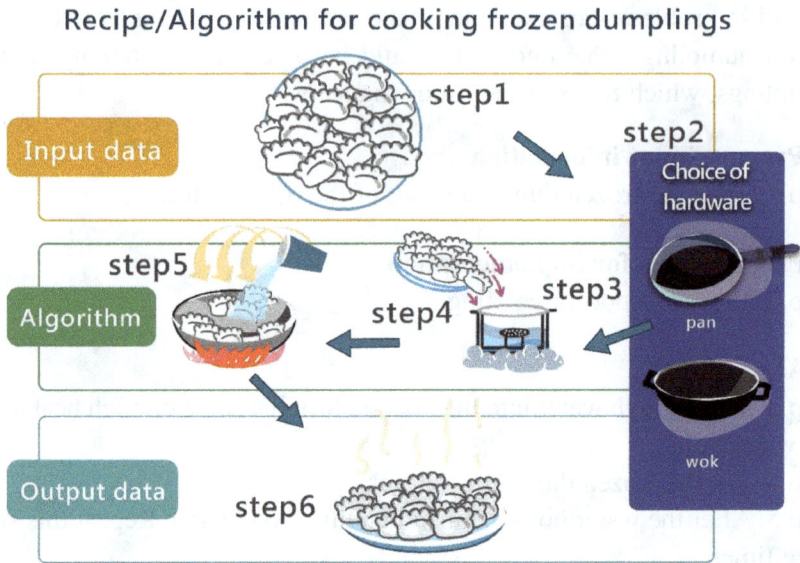

Figure 6.1. Algorithm (recipe) for cooking frozen dumplings and the choice of hardware (cookware) to implement the algorithm.

not understand why this repeated step is necessary. This is where the essence of the algorithm lies: It allows the public to achieve perfect results even when they only "know the hows but not the whys." Of course, any programmer can still have his own characteristics, just like every chef can still revise the procedures and add some additional contents on the original recipe for a better and more unique taste, and become his own special recipe.

6.2 Quantum algorithms

An algorithm is the logical thinking and steps in solving a problem, but the steps may also be adjusted according to the tools used. It is easy to fry a round-shaped egg in a frying pan with small round grids, but the shape of the egg in a large flat pan may often be polygonal. Classical algorithms contain a finite sequence of instructions to solve problems and are executed on classical computers. In the same way, quantum algorithms need to be executed on quantum computers. As introduced in the previous chapters, the difference between quantum computers and classical computers lies in the difference between qubits and classical bits. Quantum properties such as quantum superposition, entanglement, and quantum measurement do not exist in classical bits and classical computers. As a result, the thinking behind quantum algorithms is utterly different from that of classical algorithms. To fully leverage quantum properties, designing practical and efficient quantum algorithms has become an important topic. Currently, we already know that quantum computers have significant advantages over classical computers in some problems. For example, Shor's quantum algorithm can greatly reduce the computation time needed to factor a large integer, solving a problem that would take classical computers millions of years within a reasonable amount of time. Quantum computers are just tools, and quantum algorithms are operation steps on suitable tools. Even the best machine needs good algorithms to be truly functional. Quantum algorithms and quantum computers are like the form and spirit in Chinese philosophies — both are indispensable. One of the primary tasks for quantum computer scientists is to find the best operation steps on quantum computers for problems that classical computers struggle with. The following will be divided into three parts to introduce quantum algorithms: the Deutsch–Jozsa algorithm, hybrid quantum–classical algorithms, and algorithms on fault-tolerant quantum computers.

6.2.1 *Deutsch–Jozsa algorithm*

The problem of determining the properties of a black box (mathematically represented as the function f), which produces binary output (i.e., 0 or 1) for a single-bit input, was studied by Deutsch in 1985. The problem was later extended to the case of a black box with multiple bits input by Deutsch and Jozsa in 1992. They demonstrated how quantum computers outperformed classical computers on this problem by executing the Deutsch–Jozsa algorithm. Although no practical application of this algorithm has been found so far, the Deutsch–Jozsa algorithm is the first quantum algorithm to show an exponential speedup compared to any classical algorithm, thus opening up avenues for the development and application

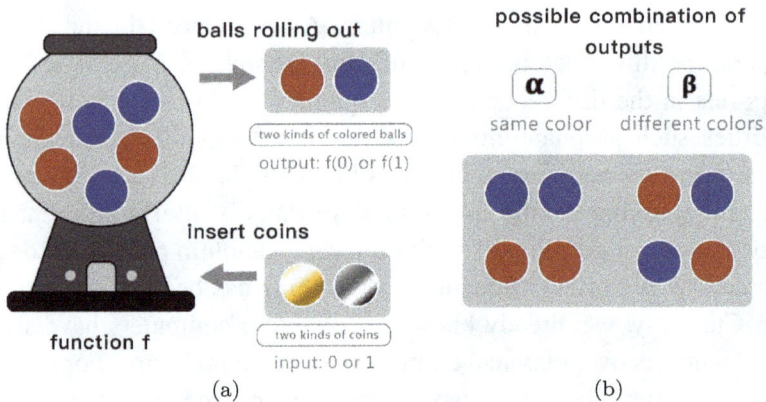

Figure 6.2. (a) Two different coins are separately inserted into a classical vending machine, which contains two colours of balls, red and blue. When a coin is inserted, a coloured ball will roll out. The vending machine is mathematically represented as a function *f*, or a mapping, where an output is obtained after each input. (b) When two different coins are inserted, there are four possible combinations of rolled-out coloured balls, which can be categorized into two types: constant states, where the balls have the same colour, or balanced states, where the balls have different colours. Classical computing requires two coin insertions to determine the combination of coloured balls rolled out by the vending machine. By contrast, Deutsch's quantum algorithm determines the combination with just one coin measured.

of other quantum algorithms. Here, we explain the Deutsch–Jozsa algorithm in an intuitive way without using mathematical language.

The mathematical problem studied by Deutsch can be understood using the classical colour ball vending machine (i.e., the *f* function) in Figure 6.2. As shown in the figure, a ball will roll out when a coin is inserted into the vending machine. Suppose we have a gold coin (i.e., 0) and a silver coin (i.e., 1) in our hand; the colour of the ball that rolls out of the machine after inserting a coin could be red or blue (i.e., $f(x)$ for $x = 0,1$). Therefore, after a gold coin and a silver coin are put into the vending machine, there are two possible situations: (α) both coins get balls of the same colour or (β) each coin gets balls of different colours, respectively. For classical computers, only by inserting both coins into the vending machine can they determine whether the balls rolled out are in the state of (α) or (β). Mathematically, state (α) is called constant (many-to-one mapping) and state (β) is called balanced (one-to-one mapping).

Deutsch cleverly utilized quantum superposition and entanglement of quantum mechanics to solve the problem by just measuring one quantum coin. The steps of Deutsch's algorithm are as follows. We will not attempt to prove the mathematics here but simply describe the logic of the algorithm: Before Deutsch inserts the coin into the vending machine, he first operates on the gold and silver coin in his hand using the Hadamard gate (H gate) mentioned in Chapter 5. The H gate transforms the gold and silver coin into a "quantum coin" with a superposition state. The function of the H gate in the quantum circuit can be imagined as a tool to modify classical coins. When this tool is used on classical coins, the classical gold and silver coin will become quantum coins that are a combination of gold and silver coin. There are two characteristics of quantum coins: (1) Once the quantum coin is measured, each quantum coin has a 50% chance of being found to be a classical gold coin or a classical silver coin. (2) When the H gate is applied on the quantum coin again, the quantum coin will return to a classical gold or silver coin.

Deutsch further transformed the classical vending machine (f) into a special quantum vending machine (U_f). The quantum vending machine has a built-in CNOT gate, which can create entanglement. As shown in Figure 6.3, when a quantum coin is put into the quantum vending machine, intuition tells us that, since the quantum coin is in a superposition state of a classical gold coin and a classical silver coin, the rolled-out colour ball should also be a superposition state combining the outcome of inserting a classical gold coin and the outcome of inserting a classical silver coin. In the following, we call the quantum coin generated by applying the H gate to the classical gold coin a quantum gold coin and the quantum coin generated by applying the H gate to the classical silver coin a quantum silver coin. When a quantum gold coin and a quantum silver coin pass through U_f, the machine U_f will output a quantum coin and an entangled result of the colour ball superposition state and the quantum silver coin, which is $y \oplus f(x)$ shown in Figure 6.3(a). After applying the final H gate on the quantum coin, the quantum coin will return to a classical gold coin or classical silver coin due to the special properties of the H gate. If the result we measured is a classical gold coin, putting classical gold coin and classical silver coin into a classical vending machine shown in Figure 6.2 will both get red (blue) balls, which is the constant state (α); on the contrary, if the

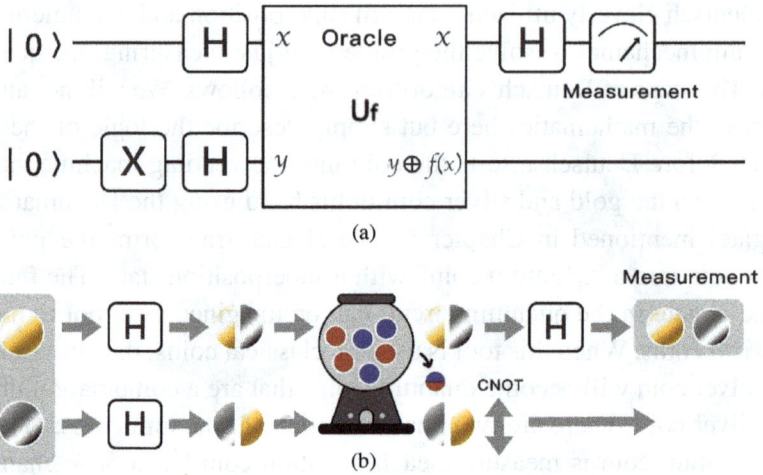

Figure 6.3. The function of Deutsch's quantum vending machine U_f is to output $y \oplus f(x)$. U_f is an example of the quantum oracle mentioned in Chapter 5. (a) Quantum circuit of Deutsch's algorithm, which enables us to determine the properties of the combined state of two colour balls rolling out of a classical vending machine by measuring only one quantum coin, killing two birds with one stone. (b) The transformation of the gold coin and silver coin in the quantum circuit. The function of the CNOT gate in the quantum vending machine is similar to the XOR gate in classical logic circuits.

measured result is a classical silver coin, we will get different colour balls after inserting a classical gold coin and classical silver coin into the classical vending machine, which is the balanced state (β). Deutsch's algorithm enables us to determine the properties of the combined state of two colour balls rolling out of a vending machine by measuring only one quantum coin.

Deutsch used the entanglement property of quantum mechanics to create a brand-new quantum vending machine in Hilbert space. The information (constant or balanced) of the colour ball combination rolling out of the classical vending machine can be immediately determined by measuring a single quantum coin rolling out of the quantum vending machine. This phenomenon where the corresponding ball rolls out after a coin is put into the vending machine is called mapping in mathematics. Clearly, once we have the appropriate quantum vending machine (U_f), Deutsch's algorithm can determine the properties of the mapping (i.e., the f function) using fewer calculations than classical algorithms. Compared to the classical method of rolling out one colour ball by inserting one coin at a time, Deutsch's algorithm exploits quantum coins with superposition states and an internally entangled quantum vending machine (U_f) to kill two birds with one stone. The Deutsch–Jozsa algorithm extends this method to the case of N coins, becoming a more powerful exponential speedup algorithm that can kill N birds with one stone.

6.2.2 *Hybrid quantum–classical algorithms*

The 1930s and 1940s were when computer technology was in its infancy. At that time, classical computers were bulky and had limited computing power. Current quantum computers have similar shortcomings. Quantum computers currently only have 433 physical qubits for IBM Osprey and will be over 1,000 qubits soon. The scale of problems they can handle is still small, and quantum error correction is not available yet. Despite these limitations, current small-scale quantum computers can still be used in conjunction with classical computers to perform valuable computations. By letting quantum computers and classical computers each handle tasks where they have the advantage, quantum and classical computers can complement each other's weaknesses. Quantum information in current quantum computers dissipates over time, causing errors to accumulate, so quantum computations with a large number of operations cannot be performed. On the other hand, while classical computers have low error rates and mature error correction techniques, they are limited by the classical nature of bits and cannot efficiently solve large-scale problems with massive variables. Therefore, hybrid quantum–classical algorithms that combine the resources of classical computers and quantum computers are the mainstream method in the near-term NISQ era. The well-known Variational Quantum Eigensolver (VQE) belongs to this kind of algorithm.

Any simple chemical reaction inside a biological organism is too large in scale for classical computers to simulate accurately. Therefore, using quantum computers for quantum chemistry simulations is the obvious choice. In quantum mechanics, a physical system is described by an operator called Hamiltonian, which contains all the information about the system, including kinetic energy, potential energy, and interactions between different particles. When all electrons within a molecule are in their lowest possible energy state, the molecule is in the most stable state, but if the molecule absorbs external energy (such as electromagnetic waves), the electrons in the molecule will jump to higher energy states (excited states), forming a less stable state. Calculating the energy state of various atoms and molecules can help scientists understand their properties and find the pathway of related chemical reactions, greatly shortening the time for drug discovery and pharmaceutical production. VQE is a specialized algorithm designed to handle these types of problems.

The architecture of VQE contains both a quantum computer and a classical computer (Figure 6.4). The quantum computer is used for preparing and calculating quantum states, while the classical computer is responsible for optimizing parameters in the quantum states. To execute VQE for quantum chemistry problems, the system's Hamiltonian must first be written in second quantization form. As previously mentioned, the Hamiltonian includes all the interactions between each electron and each atomic nucleus. However, it is almost impossible to solve for the exact wave function of the electrons and nuclei, even for a small molecule, since the different electrons and nuclei in a molecule are all intertwined by interactions. Therefore, in practice, some approximation methods are used

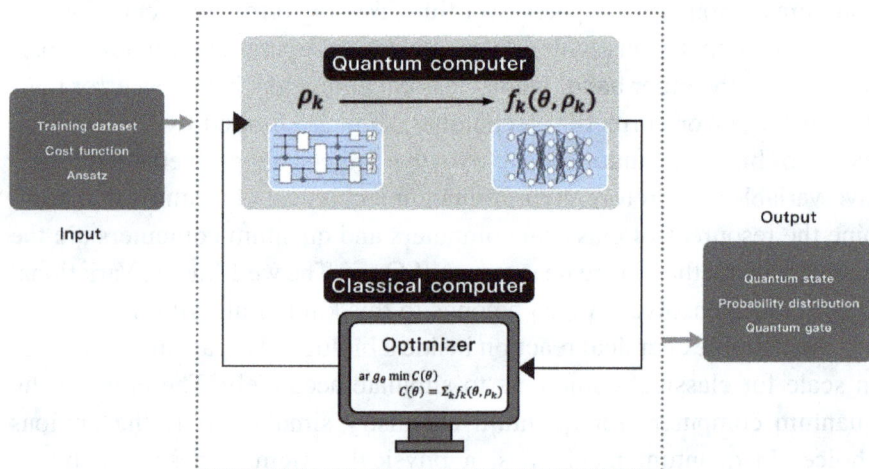

Figure 6.4. Architecture of hybrid quantum–classical algorithms.

to simplify the system's Hamiltonian to make the problem tractable. The workflow of VQE is as follows: First, a quantum circuit is constructed using quantum gates to create an ansatz (trial) state for the molecular system. The ansatz contains adjustable parameters, which can be optimized by a classical computer. Next, the energy of the ansatz state is calculated by the quantum computer. Then, the parameters of the ansatz state are optimized using a classical computer, and the new parameters found by classical optimizing algorithm are fed back to the quantum computer. This loop is repeated until the optimized (lowest) trial energy is obtained. The variational quantum algorithm is essentially a process of using a classical optimizer to train a quantum circuit. The process of obtaining model parameters through a classical optimizer is quite similar to machine learning. Starting from an ansatz state, the parameters are adjusted until the best result is produced.

6.2.3 *Algorithms on fault-tolerant quantum computers*

6.2.3.1 *Shor's algorithm*

With the rise of communication technology and the Internet, information sharing is happening around the world at all times, greatly promoting the convenience of human life. For example, during the pandemic, we could remotely access a company's computers to do our work from home with the help of encryption software. Likewise, when we want to purchase food and goods, we can easily place orders from home and have the goods delivered by simply authenticating and paying online through the Internet. However, all of these services require secure communication measures to ensure that personal information is not stolen. Thus, various encryption techniques have emerged. Currently, one of the most difficult-to-crack encryption techniques is the RSA encryption algorithm.

Current encryption systems fall into two categories: symmetric encryption and asymmetric encryption (Figure 6.5). Symmetric encryption involves sharing a single secret key between two parties. Before transmitting a file, the secret key is used to encrypt the content to ensure its security during transmission. When the recipient receives the encrypted file, they decrypt it with the same secret key. Alternatively, asymmetric

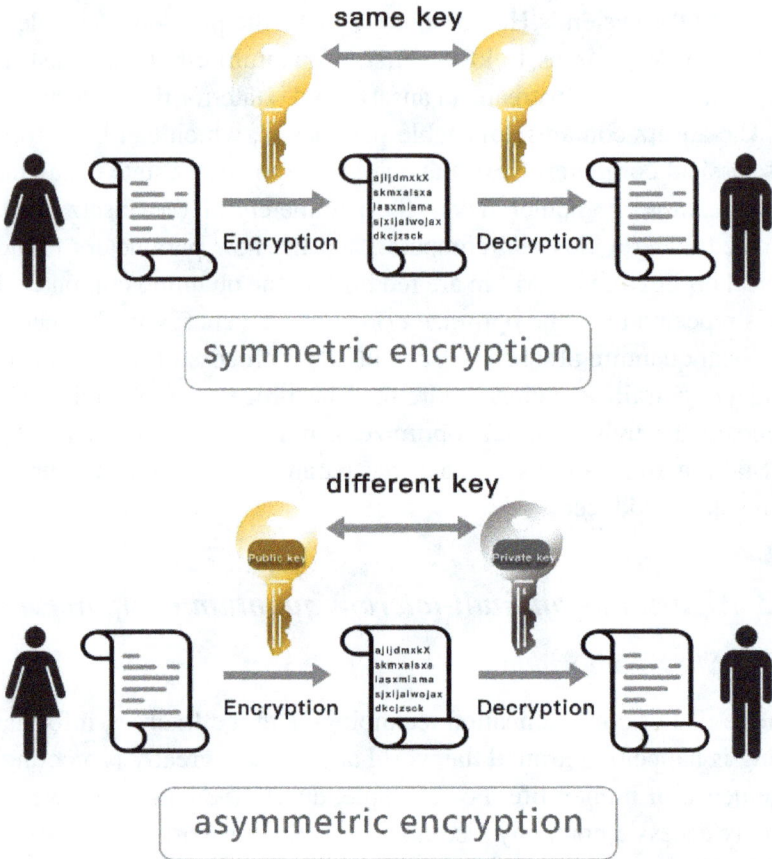

Figure 6.5. Schematic representation of symmetric and asymmetric encryption systems.

encryption can also be employed between the sender and the receiver, in which different keys are used for the encryption and decryption processes, hence the name "asymmetric." In asymmetric encryption, a pair of keys is prepared, one public and one private. The public key is used for encryption and can be made public. Anyone who knows the public key can encrypt their messages. On the other hand, the corresponding private key, which is used for decryption, must be properly kept by the individual. Symmetric encryption and asymmetric encryption have their advantages and disadvantages. The storage and transmission of the secret key in symmetric encryption are crucial. The channel for transmitting the key between the two parties must be secure; otherwise, the key may be stolen by outsiders. Although asymmetric encryption does not have the security problem of key transmission, it involves complex mathematical calculations during the encryption and decryption processes, making it slower than symmetric encryption. In actual situations, encryption systems usually combine both symmetric and asymmetric encryption.

RSA encryption is an asymmetric encryption technique, and its reliability comes from the fact that factoring large integers is very difficult. However, the advent of Shor's algorithm makes communication under RSA encryption insecure. In 1994, Shor proposed a quantum algorithm for factoring integers, which has exponential speedup over classical algorithms. Shor proved that quantum computers could effectively speed up prime factorization, enabling problems to be solved in seconds with a quantum computer that would otherwise take over 100 years to solve on a classical computer. Shor's algorithm consists of two main parts: The first part involves using a classical computer to convert the prime factorization problem into a problem of finding periods (period-finding subroutine) and the second part involves using a quantum computer to perform the calculation of finding the period. The core of the period-finding algorithm is the Quantum Fourier Transform (QFT), which is essentially a discrete Fourier transform in mathematics. We can intuitively understand the concept of Fourier transform through Figure 6.6: The waveform in space can be regarded as the combination of sine waves of different proportions, and this waveform can also be described in frequency space. As shown Figure 6.6, a complicated red wave can be decomposed into a combination of black sine waves and purple cosine waves of different proportions, which is the core concept of the Fourier transform. Due to the wave-like nature of quantum states, the optimal result can be quickly found through constructive and destructive interference of the wave. Therefore, quantum algorithms based on QFT have a speedup compared to classical algorithms.

If we want to use Shor's algorithm to find the prime factorization of an integer $N = pq$, the algorithm can be summarized into several steps:

Step 1: Select a positive integer a smaller than N and verify if it is a factor of N. If a is a factor of N, the algorithm stops since we have found one of the factors of N, and the other factor can be obtained accordingly. However, it is unlikely to be that lucky. If a is not a factor of N, we need

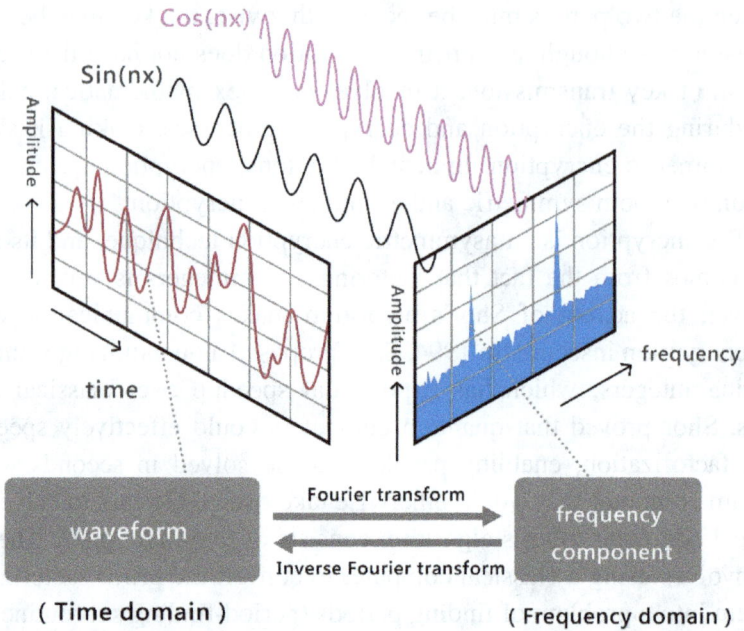

Figure 6.6. Schematic diagram of the Fourier transform. The red wave can be decomposed into a combination of black sine waves and purple cosine waves of different proportions, which is also called Fourier decomposition. The conversion of a function from the time domain to the frequency domain is called the Fourier transform, while the reverse conversion from the frequency domain to the time domain is called the inverse Fourier transform.

to find the period r of a^x mod N, where "mod N" denotes the remainder when a^x is divided by N. Notice the period r of a^x mod N is the smallest non-zero integer r such that $a^r = 1$ (mod N).

Step 2: This step of finding the period r requires the use of QFT, which is the truly quantum-accelerated process in Shor's algorithm. The QFT enables a faster determination of r that satisfies the above-mentioned equation.

Step 3: Once r is found, it means that $a^r - 1$ is an integer multiple of N, which can be expressed as follows:

$$(a^r - 1) = \left(a^{\frac{r}{2}} - 1 \right)\left(a^{\frac{r}{2}} + 1 \right) = Nd = (pq)d,$$

where d is a positive integer. $N = pq$, where p and q are the prime factors of N. If r is even, $a^{\frac{r}{2}}$ is an integer. Therefore, at this point, there are several possibilities:

I. $\left(a^{\frac{r}{2}} - 1 \right) = d, \left(a^{\frac{r}{2}} + 1 \right) = N.$

II. $\left(a^{\frac{r}{2}} - 1 \right) = p, \left(a^{\frac{r}{2}} + 1 \right) = qd$ or $\left(a^{\frac{r}{2}} - 1 \right) = q, \left(a^{\frac{r}{2}} + 1 \right) = pd.$

III. $\left(a^{\frac{r}{2}} - 1 \right) = pd, \left(a^{\frac{r}{2}} + 1 \right) = q$ or $\left(a^{\frac{r}{2}} - 1 \right) = qd, \left(a^{\frac{r}{2}} + 1 \right) = p.$

If both $(a^{\frac{r}{2}} - 1)$ and $(a^{\frac{r}{2}} + 1)$ are not divisible by N, which is the case of II and III, it means that $(a^{\frac{r}{2}} - 1)$ and $(a^{\frac{r}{2}} + 1)$ each contain one prime factor of N. Then, we can use the Euclidean algorithm to obtain the prime factors p and q of N.

If the value of r found in step 2 is odd or if $(a^{\frac{r}{2}} + 1) = 0 \pmod{N}$ (case I), we cannot successfully obtain the prime factorization of N. In such cases, we need to return to step 1 and choose another positive integer a, then repeat steps 1–3.

Shor's algorithm is a hybrid algorithm combining classical and quantum computation. Only step 2 of the algorithm involves quantum computation, which utilizes QFT to speed up the process of finding the period. The true computational advantage in Shor's algorithm comes from using QFT to find the unknown period. It takes relatively more time to find the period using classical algorithms due to the lack of phase degree of freedom in classical bits. The following explains how QFT searches for the period in a way ordinary people can understand.

The Earth has a 24-hour day–night cycle, but each person's biological time varies. For example, if a person lives in a dark cave and his sleep–wake rhythm repeats every 23.5 hours consistently, his circadian clock is a 23.5-hour cycle. There are several clocks with different periods hanging on the cave wall. Every morning when this person wakes up, he is confused by these clocks with different hours. After much thought, he comes up with a clever way to determine which clock correctly describes his

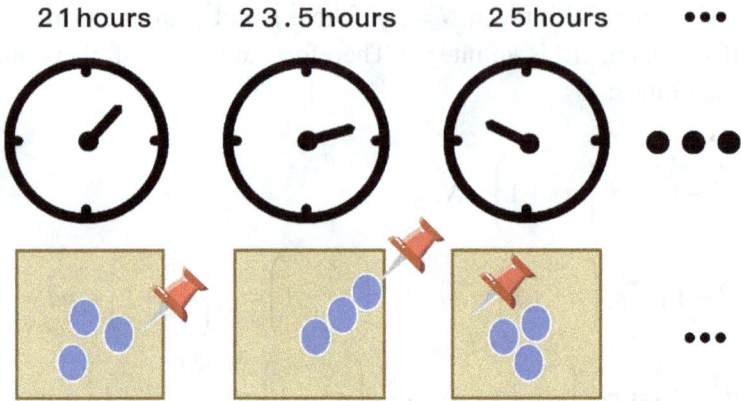

Figure 6.7.   In the cave, many clocks are on the wall, but only one clock has the same period as the person's circadian cycle. How to find the correct clock? By moving thumbtacks on note papers one centimetre in the direction of the hour hand of each clock every morning, a person can quickly find that the clock with a period of 23.5 hours correctly describes the circadian cycle. The reason is that only the path created by the thumbtack corresponding to the clock with the same period will exhibit constructive interference.

circadian cycle. He places a piece of note paper under each clock and sticks a thumbtack in the centre of the note paper. Every morning after he wakes up, he moves the thumbtacks on the note papers along the hour hand of each clock by one centimetre. As shown in Figure 6.7, after a few days, only the thumbtack on the paper under the clock with the same period as his circadian cycle will move consistently in the same direction and quickly move off the paper, while the thumbtacks on the papers under other clocks will move randomly on the paper and remain within the paper.

This process of finding the correct clock is basically a QFT. More precisely, QFT is a linear transformation that maps one complex vector to another complex vector. In the present case, the input vector is the position of the hour hands observed each morning and the output vector is the trajectory of the thumbtacks on the note paper. Therefore, the relationship between the trajectories on the note papers under different clocks and the positions of the hour hands is a linear transformation. This linear transformation maps a sequence of hour hand positions to the period of the hour hand. To understand why thumbtacks under clocks with periods different from the circadian cycle cannot leave the note paper, we must mention phase interference, a crucial difference between qubits and classical bits. In quantum mechanics, the probability amplitudes of quantum states can be positive, negative, or even complex. Because of this, in QFT, the amplitudes of quantum states that differ from the specific answer will vanish through destructive interference. In the above-mentioned example, only the hour hand that has the period of circadian cycle, for which the linear transformation points in the same direction each time, will exhibit constructive interference. Interference is exactly what happens in Shor's algorithm. The point is that the influences of other periods, which are different from the true period, are eliminated due to destructive interference. As a result, we can easily determine the true period after the final measurement.

6.2.3.2 *Grover's algorithm*

Finally, we introduce a well-known quantum search algorithm — Grover's algorithm — which enables efficient search of unsorted data. Imagine that there are n books on a table, and the book you are looking for is among them. However, you can only open one book at a time to check. How many attempts would it take to find the desired book? Using a classical algorithm, you would need to check up to n times, with an average of approximately $n/2$ attempts to find the book, resulting in a complexity of $O(n)$. However, with the power of quantum computing, Grover's algorithm allows you to find the desired book with a complexity of $O(\sqrt{n})$. That is to say, the number of operations required for searching is reduced from $n/2$ to $\sqrt{n}$. This speedup becomes more significant in systems with larger n. For example, for a system with $n = 1$ million, a classical search would require an average of 500,000 trial-and-error attempts, while Grover's algorithm would only require 1,000 operations. Besides the database search problem, there are many other problems that can be sped up by quantum search algorithms, such as graph colouring problems, store location selection, and various optimization problems in real life.

Harnessing the quantum property of superposition, Grover's algorithm considers all possibilities simultaneously in a parallelized manner, and a series of amplitude amplification steps are used to increase the probability of obtaining the quantum states corresponding to the correct answers. In Grover's algorithm, a quantum oracle that marks the correct answer is the crucial component in amplitude amplification steps. As mentioned in Chapter 5, a quantum oracle consists of a sequence of quantum gate operations. Designing a problem-specific quantum oracle that recognizes optimal solutions is necessary to implement Grover's algorithm. To provide readers with a more comprehensive understanding, the following introduces the procedure of Grover's algorithm in an intuitive and graphical manner:

Step 1: A series of Hadamard (H) gates are applied to the quantum circuit to prepare the initial superposition state $|\psi\rangle$. As shown in the right-side image of Figure 6.8, all possible solutions are simultaneously considered in the initial state $|\psi\rangle$. In the left-side image of Figure 6.8, $|w\rangle$ represents the vector of the correct solution, while $|w'\rangle$ represents the vector of non-solutions. Since the initial state $|\psi\rangle$ is composed of both the correct solution $|w\rangle$ and non-solutions $|w'\rangle$, $|\psi\rangle$ has an angular deviation θ from the non-solution axis $|w'\rangle$ in the left figure. The red bar in the right-side figure represents the probability amplitude of the quantum state ($|w\rangle$) corresponding to the correct solution. At this stage, since the initial state $|\psi\rangle$

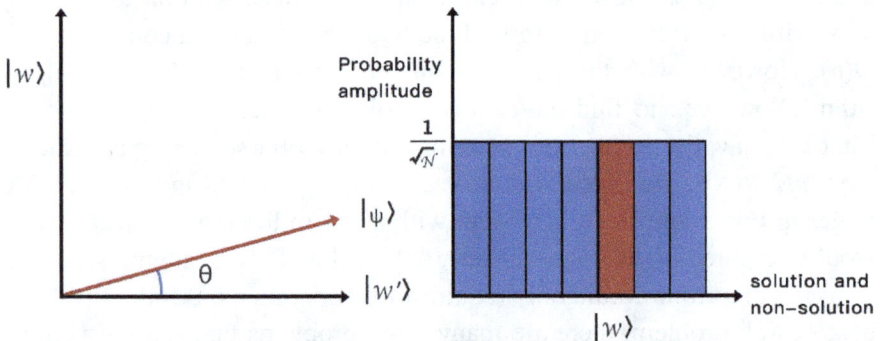

Figure 6.8. The system's state vector and probability amplitudes at step 1 of Grover's algorithm. The red bar in the right figure represents the probability amplitude of the quantum state corresponding to the correct solution, which cannot be distinguished from other states. In the left figure, $|w\rangle$ represents the vector of the correct solution, while $|w'\rangle$ represents the vector of non-solutions.

created by H gates is an equal superposition state, in which the probability of measuring $|w\rangle$ is equal to that of any other possible states, the correct solution $|w\rangle$ cannot be distinguished from other states.

Step 2: In this step, the quantum oracle is applied to the superposition state to mark the answer. The quantum oracle marks the correct solution by adding a negative sign to the probability amplitude of the quantum state corresponding to the correct solution. To implement this functionality, the quantum programmer needs to design a specific quantum oracle according to the problem. As shown in Figure 6.9, when the sign of the probability amplitude of the quantum state ($|w\rangle$) corresponding to the correct solution is reversed due to the action of the oracle, the vector $|\psi\rangle$ in the left figure will undergo a reflection about the $|w'\rangle$ axis. However, since the probability of obtaining a quantum state in quantum measurement is the square of its probability amplitude, the correct solution $|w\rangle$ still cannot be distinguished from other states at this stage.

Step 3: After marking the answer using the quantum oracle, the next step is to reflect the marked quantum state (represented by the red dashed

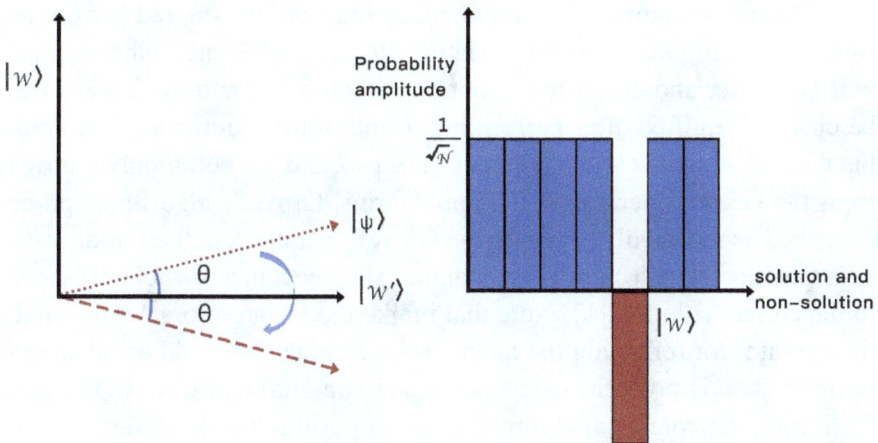

Figure 6.9. The system's state vector and probability amplitudes at step 2 of Grover's algorithm. The quantum oracle adds a negative sign to the probability amplitude of the quantum state corresponding to the correct solution.

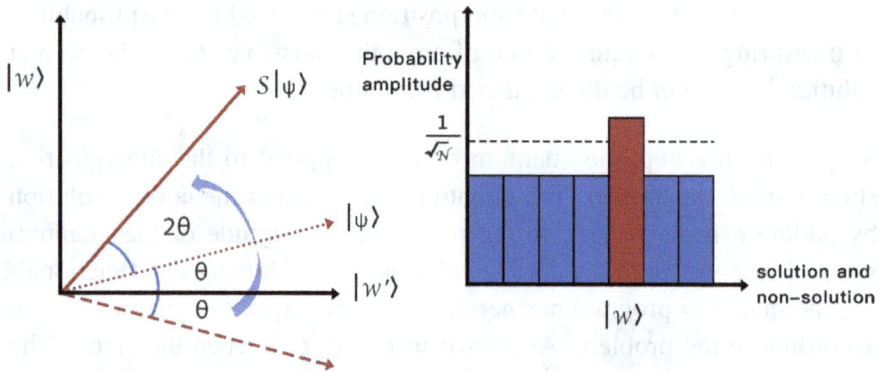

Figure 6.10. The system's state vector and probability amplitudes at step 3 of Grover's algorithm. When the marked state vector in step 2, denoted by the dashed arrow below the $|w'\rangle$ axis in the left figure, is reflected back about the initial state $|\psi\rangle$, the probability amplitude of the quantum state corresponding to the correct solution becomes significantly amplified, as indicated by the red bar in the right figure.

arrow below the $|w'\rangle$ axis) back with the initial state $|\psi\rangle$ as the axis of reflection. As shown in Figure 6.10, the quantum state at this point is closer to the correct solution $|w\rangle$ than the initial state $|\psi\rangle$. As a result, the probability amplitude of the quantum state ($|w\rangle$) corresponding to the correct solution is significantly amplified, as indicated by the red bar on the right side of Figure 6.10. By repeating steps 2 and 3, the quantum state will get closer and closer to the correct solution $|w\rangle$, and the answer can be clearly identified after performing quantum measurements. The combination of two reflection operations in steps 2 and 3 is commonly referred to as the Grover operator S. In simple terms, Grover's algorithm can be described as repeatedly applying the Grover operator S to the initial equal superposition state $|\psi\rangle$ until the quantum state becomes sufficiently close to the correct solution $|w\rangle$. Note that unlike the quantum oracle in step 2, the operator for reflecting the quantum state about the initial equal superposition state is problem-independent and straightforward to implement. Therefore, the core of applying Grover's algorithm lies in designing and implementing the first part of the Grover operator, the quantum oracle.

The approach of Grover's algorithm is fundamentally different from that of classical search algorithms. In classical algorithms, the search is based on the trial-and-error rule, resulting in an $O(n)$ complexity. On the other hand, the search in Grover's algorithm is wholly based on quantum logic. As can be observed from the process depicted in Figures 6.8–6.10, quantum search involves a series of specific reflection transformations that amplify the probability of finding the correct answer. This is a completely different logic than just trial and error, allowing for a significant reduction in search time.

The development of quantum algorithms is currently in the ascendant and in full swing, and is being studied very intensively. The above-mentioned introduction covers only the two most well-known quantum algorithms. Table 6.1 outlines several famous quantum algorithms and their potential applications. Interested readers can refer to the appropriate references for in-depth studies. In addition, there is extensive research in computational complexity comparing quantum algorithms and classical algorithms. Readers interested can refer to specialized books for further study.

Table 6.1. Some famous quantum algorithms and their possible applications.

Core component	Algorithm	Application	Advantage over classical algorithms	Potential application topics
QFT	Shor's algorithm	Cracking RSA encryption	exponential speedup	Cryptography Phase estimation
	HHL algorithm	Finding inverse matrix	exponential speedup	Hidden subgroup problem Machine learning
Grover operator	Grover's algorithm	Search problem	quadratic speedup	Search of unsorted data
Hybrid quantum–classical algorithms	VQE	Eigenvalue problem	exponential speedup when applied to chemical problems	Drug discovery Financial problems Optimization problems
	QAOA	Max-cut problem		
Quantum annealing	Quantum annealing algorithm (QAA)	Eigenvalue problem	efficiently solve the minimum eigenvalue of large matrices	Machine learning Logistics planning Financial problems Optimization problems

6.3 Concluding remarks

Since the emergence of quantum computing, there has been a significant shift in the mindset of processing data, primarily due to the transformation of programming logic caused by quantum phenomena. Quantum superposition allows data points to possess multiple possible values simultaneously, while quantum entanglement enables special correlations among the data points. Additionally, quantum interference allows the probability amplitudes of multiple data values to interfere with each other, potentially reducing computational steps in some instances. With the ingenuity of algorithm designers, significant savings in computation time and storage space can be achieved by harnessing the interplay and entanglement among quantum data. The emergence of Deutsch's algorithm has made it undeniable that quantum algorithms possess advantages. Since then, the digital world of 1s and 0s has deconstructed and evolved into the realm of peculiar quantum logic, further captivating experts' interest in developing more complex quantum algorithms, such as Shor's algorithm. Based on QFT, Shor's algorithm offers a fascinating exponential speedup over classical algorithms for the problem of large integer factorization. However, this does not imply that quantum computers can outperform classical computers in all computational problems. Currently, quantum computation is in the early stages of development, and there is still ongoing research to determine which algorithms are suitable for quantum computers and which are better suited for classical computers. The advantages of quantum computers can only be leveraged when dealing with problems that are suitable for quantum algorithms. A good quantum algorithm utilizes quantum properties to eliminate the probability of obtaining incorrect answers and enhance that of getting correct ones, which is the essence of Grover's algorithm. The future direction of quantum algorithm development is to leave digital computation to classical computers and assign the computation that harnesses quantum properties to quantum computers, combining quantum algorithms with classical algorithms to form hybrid algorithms.

Quantum algorithms indeed have an advantage in solving some problems that are beyond the capabilities of classical algorithms. Even the fastest classical computers operate with a sequence of alternating

1s and 0s, whereas quantum computers leverage the properties of Hilbert space. Quantum computers can harness quantum properties to simulate the behaviour of the universe in multi-dimensional space, leading to a vibrant and extraordinary new chapter in the realm of quantum computing, filled with excitement and limitless possibilities. However, one cannot expect quantum computers to possess superhuman powers to instantly solve all the world's difficult problems. Current quantum computers still have their limitations. While quantum computers will break certain existing classical cryptographic methods in the future, new quantum-resistant, post-quantum cryptographic techniques will also be developed. This complementarity between competition and cooperation is the driving force behind technological progress and fuels the ambition to advance both quantum computing and classical computing. Once the perfect spear appears, the perfect shield will soon follow, as it is the fundamental law of the evolution of nature. Quantum computers and classical computers will both grow in the future due to their competition, and humans will also further benefit from the integrated applications of quantum computers and classical computers.

Chapter 7

Quantum Machine Learning and its Applications

Turning quantum artificial intelligence and quantum deep learning into practical tools relies on the in-depth integration of fundamental science and mathematics.

— Shing-Tung Yau (1949–)

Nature isn't classical, dammit, and if you want to make a simulation of nature, you'd better make it quantum mechanical.

— Richard Feynman (1918–1988)

7.1 Introduction

Supercomputers are rapidly advancing, exemplified by the world's fastest Fujitsu supercomputer, "Frontier," capable of performing calculations at the rate of 10^{18} floating-point operations per second (exaFLOPS). The progress in supercomputers has led to increasingly accurate calculations for many existing complex problems, including weather forecasting and disaster prevention. However, classical computers still face challenges when it comes to simulating real material systems with a large number of electrons. While classical computers can simulate some two-level quantum systems, when dealing with problems involving n two-level quantum systems, the power of exponential growth (2^n) renders the computation of real material systems virtually impossible. Suppose there is a multi-electron system with 2^n possible electronic configurations. If n is 20, storing all possible configurations with today's computer memory would require about 128 KB of memory, which is manageable for most computers. However, when $n = 60$, it would require 128 petabytes, which is equivalent to 131,072 TB (1 PB = 1024 TB). This number far exceeds the memory capacity of the most powerful computers in the world. When the number of electrons in a system reaches several hundred, the memory required to simulate this system will exceed the total number of particles in the entire universe. Moreover, numerous 2^n quantum states will also affect each other, much like the mutual interference caused by the continuous influx of waves during the tide of the sea. Therefore, to understand the evolution of quantum states in material systems, we must track the dynamical changes in a superposition of 2^n possible quantum states, making the task extremely complex. Unfortunately, material systems with hundreds of electron systems are very common in our lives, and almost all materials and drugs have at least hundreds of electrons. This implies that the current design structure of classical computers can never simulate the quantum dynamics of any real system, making it impossible to use classical computers to develop new drugs and materials. Furthermore, with the proliferation of artificial intelligence and deep learning, it has also become evident that the computational power, bandwidth, and memory space of classical computers are insufficient for handling real and immensely complex systems. Quantum computing, on the other hand, offers advantages in at least the following three major aspects, providing opportunities for dealing with complex systems:

(1) **Computational power advantage:** Quantum computing leverages superposition and entangled states to achieve quantum parallelism, providing a computational advantage. By combining superposition, entanglement, and the unique quantum interference of quantum states, quantum computing can even provide exponential speedup on specific problems.

(2) **Storage advantage:** The capacity of quantum state space (complex Hilbert space) grows exponentially. As mentioned earlier, a quantum computer with 60 qubits already surpasses the memory capacity of the

most powerful classical computers, potentially saving a significant amount of physical storage space.

(3) **Bandwidth advantage:** Quantum data can be compressed for storage, with the potential for bandwidth optimization in the future.

In the 1980s, Richard Feynman published "Simulating Physics with Computers," emphasizing that only quantum computers could simulate complex quantum systems. However, for many years, the concept of quantum computing remained largely theoretical due to the absence of systems with a sufficient number of qubits. It was not until recent years when quantum computer hardware made rapid advancements. In 2016, IBM introduced the first quantum computer with 5 qubits accessible via the cloud and launched the open-source framework Qiskit. In 2022, IBM introduced the Eagle quantum chip with 127 qubits and, in 2023, Osprey with 433 qubits. IBM stated that simulating Eagle would require more classical bits than the number of atoms in every person on Earth combined. In other words, classical supercomputers can no longer simulate the total state space of this quantum chip. As discussed in Chapter 5, we are currently in the Noisy Intermediate-Scale Quantum (NISQ) era, where quantum computers still suffer from decoherence and lack the ability to execute error correction (Figure 7.1). However, IBM quantum roadmap also announced that Starling, a 100 million-qubit error-correction quantum computer will appear after 2029. The roadmap released by China's Origin Quantum also suggests that developing a fault-tolerant universal quantum computer is not far off. From 2016 to 2021, more than 400,000 users have accessed IBM's cloud-based quantum computing services, collectively executing over a trillion quantum computing instructions. Currently, nearly 2 billion instructions are executed on quantum computers daily to perform various tasks.

As quantum computer hardware gradually matures, the development of quantum computing applications and quantum algorithm is progressing in full swing in various industries. Many quantum start-up companies have also emerged. It must be emphasized that universal quantum computers do not outperform classical computers in all spectra of computations. Only when high-dimensional space is required, and the superposition, entanglement, and other characteristics of qubits can be utilized can the

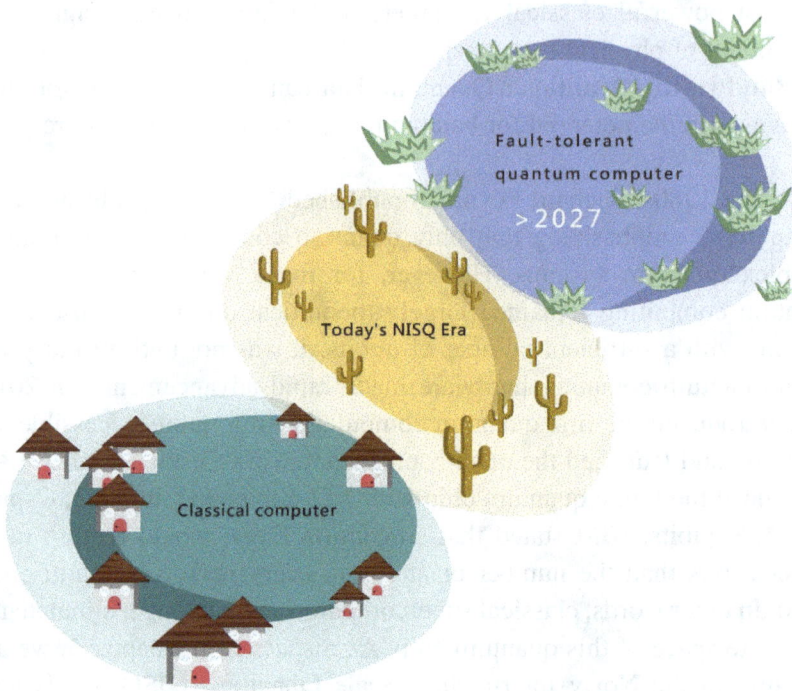

Figure 7.1.   The path from classical computers to quantum computers is filled with many challenges. Currently, we are in the oasis of NISQ, getting ready to traverse the moat of fault-tolerant universal quantum computer, and then step into the realm of true quantum computation.

value of quantum computers be truly realized. Additionally, appropriate quantum algorithms are needed to effectively harness quantum properties of quantum computers to obtain quantum advantage. Various technological approaches are being explored; they can broadly be categorized into three main areas: Quantum Annealing, NISQ Computing, and Universal Quantum Computing.

Quantum annealing is an algorithm designed to solve combinatorial optimization problems, proposed by H. Nishimori and T. Kadowaki of the Tokyo Institute of Technology in 1998. The term "annealing" in quantum annealing is derived from the annealing process used in metallurgy. During the annealing process, the material is heated to a high temperature and then gradually cooled. The material will eventually tend towards the most stable and lowest energy state. Similarly, in combinatorial optimization problems, the goal is to minimize a function associated with the real-world problem. In quantum annealing, we initialize the system state as a quantum superposition of all possible solutions. Through external control, we guide the system's state to evolve slowly towards the lowest energy state, which corresponds to the optimal solution in the combinatorial optimization problem. Compared to classical optimization algorithms, quantum annealing algorithms leverage quantum tunnelling effects, preventing the system state from being confined to local minima. This enables a faster convergence to the global minimum, making it suitable for addressing problems with many local minima. As mentioned in Chapter 5, a special-purpose quantum computer that implements quantum annealing algorithms is called a quantum annealer. Currently, the quantum annealer built by the Canadian quantum computing company D-Wave has reached several thousand qubits.

Currently, due to the limited number of qubits in quantum computers, which is only several hundred, and the noise level within each qubit, they cannot be used for error correction. Therefore, only a limited number of quantum gate operations with the current fidelity can be performed. In the presence of decoherence, completing all instructions within the coherence time of the quantum state and obtaining the computation result comprise NISQ computing. Although the number of quantum gates that can be executed is still limited by the quantum gate noise level and coherence time, quantum computers with 50–100 qubits have already outperformed classical computers for specific tasks. A new goalpost set by IBM is the 100 × 100 challenge in 2024. IBM claimed that they will be able to run a circuit with 100 qubits and a depth of 100 with three-nines of fidelity (99.9%). This 100 × 100 challenge should provide some quantum advantages in certain easy applications. There are also proposals for Perfect Intermediate-Scale Quantum (PISQ) computing, where quantum algorithms and new applications are developed based on the concept of perfect qubits and noise-free quantum gates. These quantum algorithms and applications are then evaluated on quantum computing simulators running on classical computers. The advantage of PISQ is that once a quantum computer matures in the future, the research results can be directly applied without the issue of compatibility with fault-tolerant universal quantum computers.

If a universal quantum computer with a million qubits becomes available in the future, quantum computing will be able to assist in solving a wide range of problems and will become part of mainstream usage. Companies like Google, IBM, and Honeywell have recently announced plans to launch fault-tolerant universal quantum computers between 2025 and 2030, which will help address issues in fields such as biomedicine, agriculture, and finance. Therefore, the current NISQ stage is the best time to learn quantum computing as it presents the greatest opportunities.

What kind of changes will the widespread applications of quantum computing bring to society? Figure 7.2 outlines areas where quantum computing finds application and emerging fields created by quantum technology in the "Second Quantum Revolution." In the following, we list some of the major research directions for quantum computing in both the short term and long term. The short term refers to problems that can be

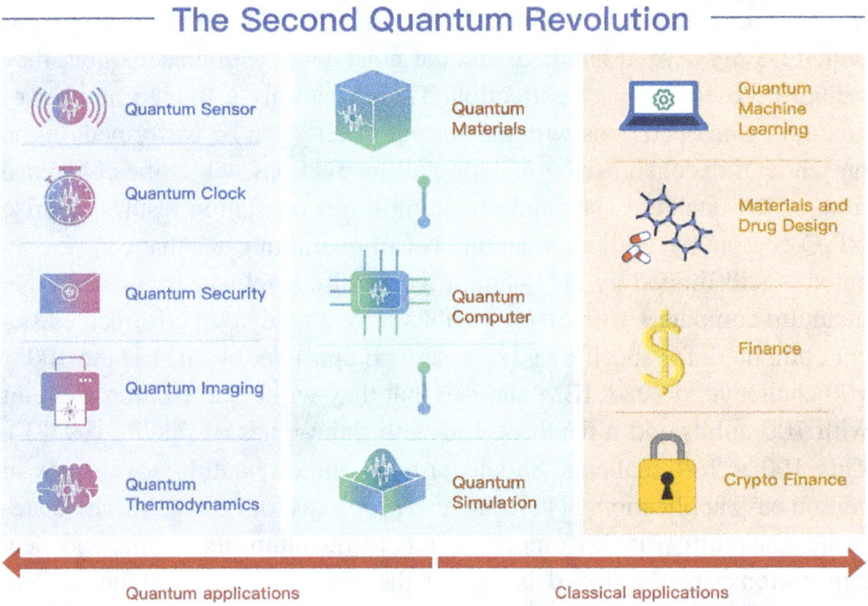

Figure 7.2. The applications of quantum computing span various domains, with some leveraging quantum advantages in existing problems, and others emerging as new fields created by quantum technology.

addressed with quantum annealing and NISQ, while the long term refers to problems that can only be tackled once fault-tolerant universal quantum computers become available.

7.2 Application fields of quantum computing

In May 2022, a research team led by the University of Innsbruck in Austria successfully demonstrated a set of fault-tolerant universal gates on two logical qubits in an ion trap quantum computer, which consisted of 16 trapped ions. Quantum information was stored in two logical qubits, with each logical qubit encoded by seven ions (physical qubits). Their result marked the first implementation of two-qubit gates on fault-tolerant qubits. In 2022, Harvard University used a quantum processor with 289 qubits to solve a combinatorial optimization problem that is challenging for classical computers. This processor featured a circuit depth of 32 layers. The research team employed a quantum-classical hybrid algorithm with a closed optimization loop, which allows for automatic feedback to optimize the parameters of quantum computations. Research results indicated a speedup in solving the problem with quantum processors, instilling strong confidence in the ability of future hybrid quantum–classical computing systems to perform better than current classical computers. In the future, quantum computers are expected to be used for simulating complex reactions in quantum many-body systems and expediting the drug development process. Moreover, quantum computers can assist in addressing fundamental scientific questions, developing new materials, solving optimization problems related to complex societal phenomena and massive data in megacities, and resolving challenges encountered in various industries. In June 2023, *Nature* published an article titled "Evidence for the utility of quantum computing before fault tolerance" by researchers from IBM Quantum and the University of California, Berkeley. IBM claimed to have entered a new era of quantum utility. They demonstrated that existing quantum computers can perform reliable calculations at scales beyond brute force classical computing methods, providing precise solutions to computational problems.

7.2.1 *Applications in chemistry, materials, and drug discovery*

In the process of exploring and researching new materials, traditional chemists often need to conduct numerous experiments to find the desired products. The process involves using different raw materials under various experimental conditions, such as temperature, pressure, acidity, and solvents, to produce different outcomes. However, whether one can obtain the desired results often depends on experience and some luck. A typical example from everyday life is Teflon, a material that was discovered accidentally by DuPont chemist Roy Plunkett while attempting to make a new refrigerant. If it is possible to first predict the chemical reaction pathways using quantum computing, it would significantly reduce unnecessary trial and error, providing substantial assistance in advancing the field of chemistry.

Modern computational chemistry methods can be roughly categorized into two main types based on the theories they use: molecular mechanics and quantum chemistry. Molecular mechanics views molecules as assemblies of atoms connected by springs (bonds) and constructs a total potential energy function corresponding to molecules based on this concept. With this potential energy function, changes in the molecular geometry are explored to find the system's lowest energy state, ultimately identifying the optimized molecular structure. However, molecular mechanics can only calculate the structure and properties of molecules; it cannot be used to investigate problems related to electronic effects, such as the process of chemical bond formation and breakage. After the development of the Schrödinger equation, scientists began attempting to use the theory of quantum mechanics to explain the structure and chemical phenomena of chemical substances. Starting from the many-electron Schrödinger equation, Douglas Rayner Hartree and Vladimir Fock developed the Hartree–Fock method. The Hartree–Fock method treats each electron as moving in the average field created by the other electrons. Taking into account Pauli's exclusion principle, they formulated the well-known Hartree–Fock equations for calculating the wave functions of multi-electron substances, laying the foundation for modern quantum chemistry. However, due to the complexity and cost of quantum chemistry calculations, it was not until

Walter Kohn and Lu Jeu Sham, among others, proposed Density Functional Theory (DFT) that it became possible to calculate large molecular systems. Still, the accuracy of the approximations adopted in DFT calculations may vary from system to system.

Although the computing power of computers continues to grow, simulating large quantum systems using classical computers is still limited. When the number of electrons in the simulated system exceeds a certain threshold, the rapidly increasing number of possible electronic configurations makes it infeasible for classical computers to handle exact quantum chemistry calculations of real multi-electron systems. Dirac once said, "The underlying physical laws necessary for the mathematical theory of a large part of physics and the whole of chemistry are thus completely known, and the difficulty is only that the exact application of these laws leads to equations much too complicated to be soluble." Richard Feynman believed from the outset that the study of matter must incorporate quantum computing because quantum phenomena are inherent to the natural world. Without the need for approximations, directly simulating the natural world using quantum systems is the most appropriate approach.

Currently, chemistry is widely considered the first field where quantum computing has the opportunity to enter practical applications. By harnessing quantum properties, quantum computers can handle computations that are challenging for classical computers, enabling the accurate simulation of complex chemical processes. For instance, catalysts for clean energy, enzymes in biological systems, new materials for solar cells, and even room-temperature superconductors could all be analyzed and simulated in advance by quantum computers in the future. As previously mentioned, when the number of electrons in a molecule exceeds about 60, certain approximation methods must be adopted for classical computers to handle the simulation. However, all of these approximation methods neglect some type of effects and cannot fully describe behaviour of electrons in materials. Therefore, approximate algorithms cannot precisely determine material properties, meaning that using them for designing new materials is almost impossible. Quantum computing, with its vast Hilbert space, can simulate complex electron–electron interactions, consequently allowing for the precise design of intricate new material structures.

In August 2020, Google research team used the Variational Quantum Eigensolver (VQE) algorithm to perform the Hartree–Fock calculation of a diazene molecule composed of two nitrogen and two hydrogen atoms on quantum processor Sycamore. In particular, they successfully simulated the chemical mechanism of isomerization of diazene. This result marked the first quantum computation of a chemical reaction pathway by a quantum computer. Although this reaction can be easily simulated using current classical computers, the VQE simulation demonstrated the feasibility of quantum computing on quantum chemistry calculations. It confirmed that the calculation results of today's quantum computer can indeed achieve the accuracy required for experimental predictions, revealing the potential for simulating more complex molecules using quantum methods. In 2017, the IBM Quantum team spent 45 days simulating the behaviour of a lithium hydride molecule using quantum computing. In 2021, with the progress of hardware and software, IBM was able to complete the same simulation in just 9 hours. Recently, Mitsubishi Chemical, JSR Corporation, Keio University, and IBM collaborated to use quantum computing in the search for the molecular structures of new OLED materials. With the rise of electric vehicles, the development and optimization of batteries have become crucial for extending their range. Leading international companies are increasingly turning to quantum computing to simulate a wide range of molecular properties and behaviours, exploring and analyzing potential battery materials. For instance, in 2021, Samsung Electronics, in collaboration with researchers from Imperial College London, used Honeywell's H1 system to explore the applications of quantum computing in battery development.

IBM is also partnering with Daimler in an effort to develop next-generation battery systems using quantum computing. Finding environmentally friendly refrigerants is a key challenge in achieving sustainable development for the future. Quantum computing company Quantinuum is working with Honeywell to research the utility of quantum computing in designing new refrigerants. Environmentally friendly refrigerants are characterized by low toxicity, low flammability, stability, low Global Warming Potential (GWP), and low Ozone Depletion Potential (ODP). Researchers in this project use the quantum computational chemistry software platform developed by Quantinuum, InQuanto, to simulate a reaction between methane gas, a simple refrigerant, and a simple atmospheric radical.

The importance and value of quantum computing in biomedicine and the pharmaceutical industry are evident. Quantum computing not only has potentially high practical value in areas such as protein folding, epigenetics, drug discovery, and drug design but also has application opportunities in DNA sequencing and genome-wide association studies. Knowing the genome sequence alone is insufficient to understand the function of proteins. Proteins can assemble into three-dimensional structures in a cellular environment, and protein folding is the process by which proteins obtain their functional forms. Protein folding can be modelled as an optimization problem, but classical computers are incapable of finding the optimal solution for such large-scale systems. Epigenetics is the study of how behaviour and environment cause changes that affect the way genes work. Unlike genetic changes, epigenetic changes are dynamically reversible and do not change your DNA sequence, but they can change the way your body reads the DNA sequence. The reversibility of quantum computing is particularly suitable for studying the impact of the environment on the functioning of genes.

As previously mentioned, quantum annealing is well suited for solving optimization problems. However, currently, the use of quantum annealing algorithms for protein folding problems can only tackle problems of small-scale systems, with the resolution of larger systems awaiting the advent of universal quantum computers. Quantum computing can also be employed in the design of targeted molecules for specific diseases through molecular dynamics simulations, which cure patients while minimizing side effects. Quantum computing can play a role in early-stage drug development, helping accelerate the simulation of interactions between specific compounds and biomolecules such as proteins and DNA/RNA. This will reduce trial-and-error time and lower research costs. It is expected that quantum computing could assist in identifying the correct biological pathways, enabling a comprehensive analysis of drug efficacy and side effects. The ultimate goal of applying quantum computing in drug development is to achieve precision medicine and personalized drugs. Menten AI has partnered with D-Wave and Rigetti Computing to incorporate AI and quantum computing into drug development, with a particular focus on drugs targeting COVID-19. The designed molecules are being tested against the virus. Menten AI is researching peptides, which are amino acid chains similar to proteins and have the potential to slow ageing, reduce inflammation, and clear pathogens from the body. There is even a concept of "pharmacy of the future," where quantum computers can quickly analyze individual health conditions and formulate personalized prescriptions, making customized drugs and health supplements.

During the critical times of a pandemic, expediting new drug development becomes imperative, and the pharmaceutical industry, with its substantial resources and research needs, has become a natural customer for quantum computing. As early as 2017, the American pharmaceutical company Biogen joined hands with Accenture and 1Qbit to use quantum computing to accelerate new drug research for Alzheimer's disease and multiple sclerosis. In October 2020, the UK's Cambridge Quantum Computing (CQC) partnered with the Dutch pharmaceutical company GlaxoSmithKline (GSK) to design algorithms to facilitate drug development. In 2021, the German pharmaceutical company Boehringer Ingelheim, in collaboration with Google Quantum AI, IBM, and the Cleveland Clinic, worked to harness the promising potential of quantum computing and artificial intelligence to develop drugs to combat viruses. Their primary objective was to achieve more precise simulations of large molecules, expediting the research of new drugs and therapies.

New materials and molecules have huge economic value, especially in the case of drug development. If quantum computing can transform drug development by replacing the traditional trial-and-error experimental methods with computational analysis, the time required for drug development can be reduced and substantial development costs can also be saved. Certainly, the practical application of quantum computing in drug development also requires the concurrent development of specialized quantum algorithms. Many start-ups are currently moving in this direction. In 2021, the Chinese company Origin Quantum released the quantum chemistry software ChemiQ 2.0. This application software has been instrumental in advancing practical quantum computing research in the fields of new medicine, new materials, and new energy, providing a starting point for development. On May 24, 2022, Quantinuum launched its quantum computational chemistry software platform, InQuanto, which enables the testing of various quantum algorithms on quantum computers.

7.2.2 *Quantum artificial intelligence and quantum machine learning*

Quantum artificial intelligence (QAI) is a promising technology that can augment the speed of classical artificial intelligence (CAI) technology and bring great changes to many fields. QAI uses the unique properties of quantum mechanical effects to enhance the capabilities of CAI. However, quantum computing is still in its NISQ stages, and there are many technical hurdles that must be overcome before the implement of QAI. On the other hand, some experiments have been conducted using quantum computing to enhance classical machine learning (CML) algorithms in a hybrid way, which is still also in a proof-of-concept stage. Many experts believe that as soon as fault-tolerant quantum computers emerge, QAI will soon be applied in decision-making, financial analysis, and language translation. However, it may take many years for QAI to transform our world.

CML has penetrated deeply into business, science, and various fields. Its main goal is to train computers to understand increasingly complex data, assisting in rapid analysis and accurate decision-making. Common applications include pattern recognition and data classification. In supervised learning, the model parameters are optimized through training data, and then these trained parameters are applied to process new data. In concrete terms, given training data $(x_1, y_1) \ldots (x_n, y_n)$, we want a machine to learn the mapping between x and y. Taking image recognition as an example, we need to show the computer a set of animal photos (x) and inform it of the corresponding animal for each photo (y). After training the model, we hope that when we present the computer with a new photo it has not seen before, the computer can accurately identify which animal is in the photo. Image recognition is a typical classification problem in machine learning. Classification algorithms in machine learning have evolved from the simplest perceptron algorithm, which could only distinguish between two classes in 1957, to the widely used deep learning based on artificial neural networks today. In deep learning methods, we need to optimize a vast number of parameters in artificial neural networks, which requires significant computational resources and time. However, if the network has too few parameters, it cannot accurately describe all the features in the data and, therefore, is unable to help us solve problems.

High-dimensional datasets, characterized by many features relative to the number of observations, present significant challenges for machine learning models with a "curse of dimensionality." In such environments, models are prone to overfitting, where they excel at memorizing the training data, including noise, but fail to generalize to new and useful data. The sparsity of data in high-dimensional spaces complicates the learning process further, as the volume of data required to represent the space adequately grows exponentially with each added dimension. However, quantum machine learning (QML) is promising for enhancing classification tasks in high-dimensional datasets. Leveraging the principles of quantum mechanics, such as superposition and entanglement, quantum computers can process and represent data in ways that are fundamentally different from classical computers and this capability efficiently handles the vast amounts of information in high-dimensional spaces.

Quantum computers can process data in high-dimensional Hilbert space, and the quantum superposition and entanglement properties of quantum computers provide quantum advantages in certain computations, enhancing the performance of machine learning. For example, current machine learning often involves extensive linear algebra operations, and the Harrow–Hassidim–Lloyd (HHL) quantum algorithm can exponentially accelerate the solution to multi-dimensional linear algebra problems compared to classical algorithms. Therefore, the combination of quantum algorithms and machine learning has given rise to QML. Quantum neural networks in QML, counterparts to classical neural networks, are typically implemented using parameterized quantum circuits, where some quantum gates applied to input quantum states contain adjustable parameters. Similar to CML, these parameters are optimized based on the problems under consideration. For example, the VQE mentioned in Chapter 6 optimizes parameters with the aim of minimizing system energy.

In machine learning, Principal Component Analysis (PCA) is often used to identify important features in the data. PCA is commonly applied in unsupervised learning scenarios where labelled data (y values) are not available. In PCA, we need to calculate the largest few eigenvalues and corresponding eigenvectors of the data's covariance matrix, and the dimension of the covariance matrix is typically large. Seth Lloyd of the Massachusetts Institute of Technology proposed quantum PCA. Employing the HHL quantum algorithm, he showed that one can perform PCA exponentially faster than classical algorithms. In quantum computing, representing an N-dimensional vector requires only $\log_2^N$ qubits. Once we encode the data into quantum states, certain operations on the data exhibit exponential acceleration compared to classical computers, such as quantum Fourier transform mentioned in Chapter 6. Even in just performing a simple inner product operation between two vectors, a quantum computer is exponentially efficient compared to a classical computer. In machine learning, the support vector machine (SVM) method is commonly used to handle binary classification problems due to its many favourable properties. One of these is the absence of the problem of getting stuck in local minima during parameter optimization. Once the optimal solution is found, it signifies the discovery of the global minimum. Another advantage is that in SVM, we only need to know the relationships between different data points. In some cases, we do not even need to know the data points itself. In SVM, the key is to calculate the kernel function between different data points, and this calculation involves inner product operations. Therefore, a quantum support vector machine (QSVM), which uses a quantum computer to process data, can result in exponential time savings compared to its classical counterpart. The team at the University of Science and Technology of China has already conducted experiments to verify the feasibility of QSVM using NMR quantum computers.

Certainly, applying quantum computing to machine learning is not without its challenges. Currently, QML faces several obstacles. First, machine learning typically deals with classical data, so classical data must be encoded into quantum states that can be input into a quantum computer. Many studies have pointed out that this process of preparing quantum states is not always efficient. Even with quantum random access memory, it is only possible to efficiently encode an N-dimensional vector into $\log_2^N$ qubits when components of the data vector are relatively uniform. On the other hand, if one component of the data vector is much larger than the others, the inefficient data encoding process can eliminate the exponential advantage of quantum speedup. In other words, achieving exponential speedup in quantum PCA and quantum SVMs requires certain conditions, such as the good nature of the data vectors under consideration. However, if we already assume some properties of the input data, classical algorithms may also become faster for specific operations on the data. Therefore, in order to maximize quantum advantages, scientists must identify specific types and properties of data that can be efficiently encoded into quantum states. Additionally, classical algorithms should not be able to quickly perform operations on such data. Second, we are currently in the NISQ era, so even if, in theory, quantum computers can perform specific operations faster than classical computers, the imperfections in hardware will lead to higher error rates compared to classical computers. The third challenge is that quantum computing operations ultimately require measurement to read out information. It is necessary to perform repeated measurements multiple times to obtain the probability distribution of a quantum state, meaning that the execution of quantum algorithms needs to be repeated. This process also has the potential to diminish quantum advantages.

Therefore, rigorously proving the superiority of QML over CML requires considering the entire process, from data input to operation to the readout of information. Recently, a team from the California Institute of Technology investigated the performance of QML models in learning the mapping of quantum origin. This mapping, to be learned, is derived from a specific quantum process, such as a chemical reaction in a laboratory experiment. This team provided rigorous mathematical proof that, under the condition of demanding small prediction errors for each data point, QML exhibits an exponential advantage over CML for specific problems. These specific problems where QML excels include predicting observables of quantum systems or finding the maximum eigenvalue of an unknown quantum state. It is worth noting that achieving these advantages necessitates the availability of quantum random access memory. In 2021, Google introduced TensorFlow Quantum, an open-source library that integrates parameterized quantum circuits with traditional neural networks. This hybrid approach aims to combine the respective strengths of quantum computers and classical computers. Reflecting on Feynman's initial motivation for suggesting quantum computers, it is conceivable that their optimal application still lies in solving problems of quantum systems. Identifying problems suitable for QML and addressing the three major challenges mentioned earlier is currently the key focus. To fully unleash the power of QML, collaborative efforts from computer scientists, mathematicians, and physicists are essential.

7.2.3 *Applications in financial industry*

Since the 1950s, computers have played a pivotal role in the financial industry. As quantum computing technology matures, Goldman Sachs, a prominent investment bank, believes that quantum computers could be applied in the financial industry within the next 5 years. Quantum computing has a natural advantage in solving high-dimensional problems. Moreover, it can prevent the system from getting stuck in local minima thanks to the quantum tunnelling effect. The current challenge lies in the fact that under the limitations of NISQ, error correction cannot yet be implemented. Quantum computing can, in principle, provide speedup compared to classical computing, helping us identify the correct direction. However, in the era of NISQ, errors in the results are unavoidable. For some specific financial applications, within a tolerable range of errors, speed and decision-making direction are far more important than errors. Therefore, leading financial enterprises, such as J. P. Morgan and Goldman Sachs, have established quantum departments to develop quantum finance applications. In China, Xinhua Finance and Origin Quantum jointly released the "Quantum Financial Application." The application is available on the Xinhua Finance mobile app, providing services such as portfolio optimization based on quantum simulation. In addition, Origin Quantum has developed QSVM, applying it to practical scenarios in the financial field, such as predicting stock amplitudes and developing multifactor stock selection models. The application of quantum computing in the financial field is extensive, aiming to reduce costs and processing times. Currently, it mainly includes risk control, derivatives pricing, portfolio optimization, arbitrage trading, and credit scoring.

Risk management is a core piece in the financial industry. Value at Risk (VaR) is a quantitative indicator for market risk, which originated from J. P. Morgan, the renowned investment bank of Wall Street. J. P. Morgan required business units to submit a one-page risk report within 15 minutes after the end of each day's trading, which estimated the company's potential losses in the next 24 hours due to market fluctuations, famously known as the "4:15 report." The meaning of VaR is as follows. If the estimated one-day 5% VaR is $10 million, it means there is a 5% probability that the company's held assets will not incur losses exceeding $10 million within the next day. Currently, the most common method for calculating VaR is Monte Carlo simulation. However, due to limited computational power, there is often a trade-off between accuracy and speed. Increasing accuracy requires increasing the number of random simulations, resulting in lengthy calculation time — which could be several days when assessing the risk of large investment portfolios. IBM recently published an article, "Quantum Risk Analysis," which uses the Quantum Amplitude Estimation (QAE) algorithm to analyze financial risk. Compared to the Monte Carlo simulation, this approach can significantly reduce computation time from days to a few hours by leveraging quantum superposition and entanglement. Caixa Bank, which has the highest number of digital customers in Spain, recently employed IBM's Opensource Qiskit for quantum financial development, conducting financial risk assessments on two asset portfolios. While classical computing requires days for this complex task, using quantum algorithms can complete the simulation within minutes.

Financial simulation is like navigating a maze. Classical computing can only take one path at a time, while quantum computing, leveraging the quantum superposition, can simultaneously traverse all paths in parallel (Figure 7.3). When playing maze games as children, once entering a maze, it is challenging to find the way out, as we can only try by choosing a path and backtracking if it is wrong. A clever mind, like a supercomputer, can remember the wrong paths and mark the routes it has taken, preventing one from getting lost again. On the other hand, a quantum computer will never be trapped in a maze because quantum superposition allows it to appear on all paths simultaneously, and quantum interference will let the optimal solution naturally emerge. This is where quantum search differs most significantly from classical search.

QAE can also be applied to the pricing of "financial derivatives," which require large computational power. "Financial derivatives" are financial products derived from underlying entities, among which options are the most widely used. The price changes of options are typically

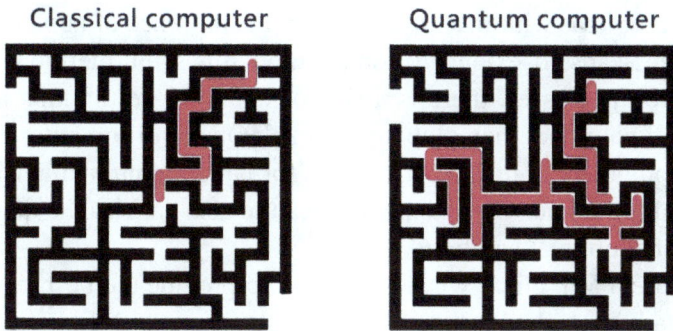

Classical computer **Quantum computer**

Figure 7.3. Schematic diagram of the difference between classical computer simulation and quantum computer simulation. Searching for the optimal solution is akin to finding a way out of a maze. Classical computers can only take one path at a time, while quantum computers can leverage superposition to process multiple paths in parallel.

simulated using Monte Carlo methods. Predictions are made about future trends after calculating the probability of specific prices. Pricing financial derivatives with current computers takes several hours, sometimes even longer. In 2020, IBM and J. P. Morgan collaborated on a study, employing QAE in option pricing. Their method can potentially speed up the pricing of options compared to traditional Monte Carlo simulations. In another study conducted by Goldman Sachs and quantum start-up QC Ware, it was estimated that quantum computing could be applied to some of the most complex computations in the financial market within the next 5 years.

Portfolio optimization is a common challenge in the financial industry, where investors seek the most profitable investments. The important question is as follows: What constitutes the optimal investment portfolio? Hedge funds strive to maximize profits, while insurance companies aim to minimize risks. According to the economics Nobel laureate Harry Markowitz, and his modern portfolio theory, investment involves choosing among returns and risks filled with "uncertainty," and the investment portfolio should be optimized with the goals of minimizing risk and maximizing return simultaneously. In the present landscape with diverse investment targets, convenient trading, rapid information dissemination, and volatile price movements, identifying the most optimal portfolio in real time remains a prominent issue in the financial field. In solving financial problems with machine learning, the presence of numerous variables often leads to extended computation times. Furthermore, during optimization processes, there is a risk of obtaining results that are only locally optimal rather than globally optimal. Quantum computer excels in parallel computation, making it suitable for handling problems with multiple variables. Its quantum tunnelling feature also enables the quick finding of global optimal solutions.

Suppose a VIP client wishes the bank to provide the best investment portfolio advice and analysis for wealth allocation, and there are 10,000 global investment targets; then, the number of possible combinations will exceed 2 to the power of 10,000! If the bank takes too much time to perform calculations, the price of the targets may have changed significantly from the prices considered in the calculation. In such cases, the calculation results are unlikely to be correct. However, if the focus is solely on speed, there is a risk of severe losses for the bank or investor due to errors in optimization calculations. Finding the optimal investment portfolio among a vast number of combinations in a short time is an urgent problem in the financial industry. Recently, the quantum start-up Multiverse Computing collaborated with the Spanish bank BBVA to employ quantum computing to optimize investment portfolios. They proposed a quantum computing method that can quickly select a portfolio of 52 assets out of 10,382 candidates. Handling such a large amount of data would take approximately two days using classical algorithms on a classical computer, but with quantum computing, results are obtained within seconds. In March 2022, QBoson, in collaboration with Everbright Technology and Beijing Institute of Quantum Information Science, released the quantum computing portfolio product "Tiangong Jingshi (天工經世) Quantum Computing Quantitative Strategy Platform." On this platform, QBoson applies coherent Ising machine (CIM) technology to solve the portfolio optimization problem based on Markowitz's portfolio theory. Being the first quantum computing quantitative strategy platform in China, "Tiangong Jingshi" aims to provide quantum solutions for financial decision-making.

Since it is possible to identify the optimal investment portfolio that aligns with low risk and high return using a quantum computer, it is reasonable to think that quantum computing can help us find potential arbitrage opportunities as well. One arbitrage strategy is cross-currency arbitrage. For instance, one can convert Euros to US Dollars, then to Japanese Yen, and finally convert back to Euros, gaining profits during this exchange rate conversion process. In 2016, the start-up company 1QBit demonstrated how to use quantum annealing algorithms to find profitable cyclic arbitrage opportunities in the foreign exchange market. UBS, one of the biggest financial services companies in the world, is also collaborating with another quantum start-up, QxBranch, to explore the application of quantum algorithms in forex trading and arbitrage. With the increasing variety and diversification of cryptocurrencies, identifying arbitrage opportunities between cryptocurrencies or even between crypto-currencies and real currencies may require the speed and accuracy that only quantum computers can provide. When quantum computing hard-ware matures, the high-speed calculations offered by powerful quantum computers could revolutionize and disrupt the current financial industry, ushering in the era of quantum finance (Figure 7.4).

Credit scoring is also a crucial issue in finance. The estimated total debt of US households is around USD 17 trillion, with a default rate of

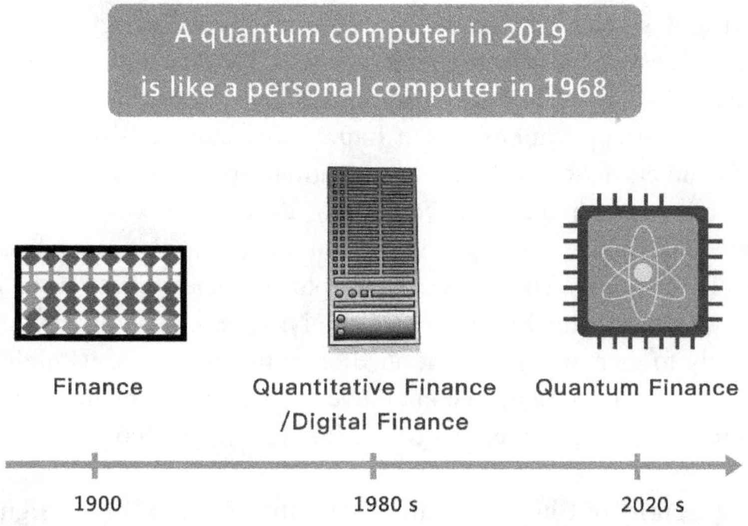

Figure 7.4. In the financial era, calculations were performed using abacuses. In the era of digital finance and quantitative finance, supercomputers were used. The era of quantum finance will be built on quantum computers.

approximately 2%. Between 2001 and 2016, 547 financial institutions in the United States went bankrupt, with non-performing loans from banks being a major contributing factor. It is important for banks and other financial institutions to assess whether a borrower has sufficient ability to repay. Before approving a loan, banks consider the borrower's income, age, financial history, collateral, and other factors, conducting credit scoring to determine the level of risk. When there is a new loan application, banks often use machine learning methods to decide which past data of the applicant provide useful information. PCA is often used to extract important features of data, simplifying data analysis and saving time. As mentioned in the previous section, PCA can be significantly sped up by using quantum algorithms, enabling a fast dimensional reduction of datasets. 1Qbit employs another approach, using the quadratic unconstrained binary optimization (QUBO) model and quantum annealing algorithms for feature selection, reducing the number of variables in machine learning calculations.

In addition to the various financial applications mentioned previously, many people are also curious about the application of quantum computing in the recent trends of cryptocurrencies and blockchain. The security of Bitcoin and similar products using blockchain technology lies in the strength of asymmetric encryption methods, making them resistant to decryption by classical computers. But, will quantum computers pose a threat to these applications in the future? The current Bitcoin protocol rewards miners based on their computational power in mining. In the future, miners with quantum computing capabilities will demonstrate computational power far surpassing that of other miners, relying solely on classical computing. The success of cryptocurrencies relies on the complexity of mathematical problems. In the NISQ era, quantum computing is unlikely to gain any advantage in Bitcoin mining due to its high error rates. However, once fault-tolerant universal quantum computers emerge, quantum computing's dominance in Bitcoin mining could become a reality.

The security of Bitcoin is built on the Elliptic Curve Digital Signature Algorithm (ECDSA). Bitcoin owners possess a private key and publish a public key, which can be easily generated from the private key but not vice versa. While it is challenging for classical computers to derive the private key from the public key, the mathematical equation of elliptic curves could potentially be broken with the advent of fault-tolerant universal quantum computers. Beyond cryptocurrencies, blockchain technology is currently applied to various emerging virtual assets, with the most recent example being Non-Fungible Tokens (NFTs). Although NFTs have yet to become popular in the mainstream, their trading activity and market acceptance are progressively increasing. Quantum computing is a threat to blockchain and has potential impacts on future financial markets. Therefore, quantum-resistant ledgers and post-quantum cryptography (PQC) are both being developed. This clash of spears and shields in the virtual world will intensify as quantum computing matures and block-chain technology becomes widespread.

7.2.4 *Applications in management of megacities and industry*

As the world's population continues to concentrate, the number of megacities with populations exceeding several tens of millions is growing. Metropolitan areas like Tokyo and Shanghai have populations exceeding ten millions. In such crowded cities, various social issues gradually emerge, making it crucial to address how to allocate resources efficiently to solve problems caused by population concentration. In this aspect, quantum computing can help address issues related to transportation and logistics. In addition, quantum computing can assist large-scale manufacturing industries in production scheduling and supply chain management, reducing costs and risks.

1. **Traffic problems:** In megacities, effectively arranging public transportation and managing traffic flow to optimize transport capacity and relieve traffic congestion are urgent problems. In 2017, one of the world's largest automobile manufacturers, Volkswagen, collaborated with D-Wave to study route planning services for 10,000 taxis in the congested city of Beijing. Using the quantum annealer of D-Wave, traffic optimization tasks can be completed in a few milliseconds. Building on their research and development experience in the United States and Germany, D-Wave and Volkswagen tested this approach in

2019 with nine buses in Lisbon, Portugal. This marked the world's first experiment of using quantum computing for traffic optimization. Through cloud-based quantum computing, the project assisted buses in determining the fastest and most efficient routes, reducing traffic congestion and travel time.

2. **Logistics:** Save-On-Foods, a retail company in Western Canada, collaborates with D-Wave, employing a hybrid quantum-classical algorithm to bring optimized solutions to its retail sales business. The outcome of the collaboration is a significant reduction in the time required for an optimization task from 25 hours to just 2 minutes, and the calculation result obtained is the true global optimum.

3. **Production scheduling:** Volkswagen employs a hybrid quantum–classical algorithm to optimize the order of car painting, significantly reducing the number of colour switches in car painting and thereby improving efficiency.

4. **Supply chain management:** DP World in Dubai, UAE, has begun to develop the application of quantum computing in the logistics and trade industry, including industrial logistics, fleet management, and transportation management. Harnessing the extraordinary power of quantum computing, DP World aims to usher in a new era of supply chain optimization.

7.3 Concluding remarks

7.3.1 *The future of quantum computing and challenges for modern countries*

If CAI wants to move into the future, increasing computing power while reducing energy consumption is a key issue. Quantum computing has powerful parallel computing potential that far exceeds classical computing in principle, and due to its reversible computing characteristics, it is also a green energy product. Quantum computing is like the Holy Grail in the field of computing power. It brings "speed" and energy saving at the same time. Quantum computing will completely change the classical computing power limitations of the current large language models (LLMs) and CAI, and will move into a new paradigm in the QAI phase. However, we are currently in the NISQ stage, with an optimistic estimate that it will progress to the era of fault-tolerant universal quantum computing in the next decade. At that time, there will be an opportunity to address problems that current computers cannot solve. The commercial cloud-based quantum computing services launched by IBM symbolize the dawn of the quantum computing era, but there is still some time before substantial enterprise applications begin. In the current stages of quantum computing, the emphasis is on solving specific problems through quantum annealing and NISQ, and practical experiences in usage have revealed substantial business opportunities. As we enter the era of fault-tolerant universal quantum computing, businesses need to identify quantum niches and governments must plan comprehensively to address quantum risks in advance. Quantum algorithms are much more efficient than classical algorithms in searching, optimization, allocation, and scheduling problems. The advantages of quantum computing have already emerged in many fields, with the first major application area being optimization problems in finance. Accenture, the world's largest management consulting and technology services provider, is developing quantum computing services and conducting a series of commercial experiments, aiming to use quantum computing for arbitrage trading, credit scoring, and transaction optimization. China's Origin Quantum has also launched quantum computing application software in chemistry and finance, and has demonstrated quantum advantages in specific problems.

7.3.2 *How enterprises and countries respond*

The quantum computer roadmap will soon exceed one million physical qubits, and quantum advantage will be reflected in all levels of applications. Quantum Computing as a Service (QCaaS) will become the future platform for general customers. QCaaS mainly develops quantum algorithms and combines them with artificial intelligence and machine learning to solve basic, industrial, biomedical, and financial problems. Due to the potential advantages of quantum computing in various fields and the possibility of overcoming the high-dimensional curse of classical computing, QCaaS is often jokingly called the saviour of classical computing, where S stands for Saviour instead of service. Since quantum computing is an important opportunity for disruptive changes in the future, many countries and emerging companies also hope to take this opportunity to change lanes and overtake others. In the future, whoever can be the first to find an effective quantum computing breakthrough will make absolutely huge profits for the industry and serve as an innovation leader to subvert the direction of industry development. Today's cell phones are millions of times more powerful than the computers used in the Apollo moon landings, and future quantum computers could be 100 million times faster than today's supercomputers! The compelling attraction of quantum computing lies in its ability to instantly solve complex computational problems that would take classical computers decades to resolve. The difference between a quantum computer and a supercomputer is even greater than the difference between a supercomputer and an abacus. It is said that the quantum computer that appeared in 2020 is similar to the mobile phone that appeared in 2010, but the impact may be more far-reaching. Once quantum computing applications really take off, they will grow on an exponential curve so fast that anyone who does not invest today will regret it tomorrow.

Quantum computers have evolved from the laboratory to become tools poised to revolutionize various industries. There is a possibility that quantum computing can help us understand biological evolution to cure diseases such as cancer and even mitigate climate change and natural disasters. Apart from IBM, Microsoft Azure Cloud, IonQ, Huawei, Google, and Amazon have also launched cloud platforms for quantum computing. Following Google's announcement of achieving quantum supremacy, IonQ officially went public with a valuation of USD 2 billion. Venture capital has actively invested in the field of quantum computing. According to an analysis by the journal *Nature*, by the beginning of 2019, private equity funds had invested in at least 52 quantum technology companies around the world. Recognizing the importance of quantum technology, governments worldwide are increasing investments in the quantum-related industry. Numerous start-ups are currently engaged in the research and development of quantum computing applications, including companies like 1Qbit, QxBranch, and Aliro, which has developed cloud-based quantum computing services and successfully completed a seed funding round. In addition to quantum start-ups, major technology companies are also actively investing in quantum technology. As quantum technology is now in a pre-commercialization phase, industry leaders such as IBM, Intel, and Google have already joined forces with many partners to address challenges in various fields by developing applications of quantum computing. For instance, ExxonMobil and IBM are working collaboratively on optimizing the transport of energy products in the global commercial fleet, aiming to use quantum computing to find the best strategies. BMW has also started using Honeywell's H1 quantum computer to optimize the scheduling of the supply chain and car assembly, seeking to increase production capacity and sales. Nippon Steel and CQC use IBM quantum computers and employ specialized algorithms developed by CQC to run simulations, discovering new types of iron crystals that classical computers cannot identify. This finding could contribute to the creation of innovative steel materials and provide insights into fundamental issues related to the Earth's iron core. Goldman Sachs and Ion Q have also demonstrated that financial computations can be processed more quickly and efficiently on quantum computers. Chinese tech giants Baidu, Alibaba, Tencent, and Huawei have long established quantum computing departments and are actively investing in quantum computing applications. Recently, JD.com has also established JD Explore Academy, focusing on quantum computing research.

Once quantum computing applications take off, they will grow exponentially. Large enterprises and countries, well aware of the logic that "not investing today means regret tomorrow," understand that the takeoff point on the exponential growth curve is now. Therefore, they are actively investing in quantum computing. Since quantum computing presents a crucial opportunity for future disruptive change, not only are the traditional powerhouses vigorously promoting quantum computing to drive the future of emerging industries but many emerging countries and enterprises are also aspiring to seize this opportunity for advancement and actively participate in the quantum race. Whoever can make breakthroughs in quantum computing to gain advantages in key industries will reap huge profits and become an innovative leader, reshaping the direction of industry development. Whether countries can make appropriate decisions during this brief opportunity of technological transformation and stand out in the global competition of the "Second Quantum Revolution" depends on the collective wisdom and the collaborative efforts of governments, academia, and industry.

Chapter 8

Quantum Metrology and Quantum Sensors

The rain of apricot blossoms can barely wet the clothes, and the wind of willows can only gently touch the face.

— Song, Zhinan (unknown)

There will be 50 billion devices, modules, sensors connected to the Internet by 2025, and that will give opportunities.

— Rajeev Suri (1967–)

8.1 Bio-quantum sensing in animals

As is well known, humans possess six senses — sight, hearing, smell, taste, touch, and proprioception. However, in the biological world, many organisms have sensory abilities that far surpass human perception and are even more sensitive than modern technological sensors. For example, we have long known that eagles can see tiny prey, ants have a keen sense of smell to detect distant food sources, and cats and dogs have hearing abilities that can pick up unusual sounds. While humans rely on technological products like the Global Positioning System (GPS) for navigation and positioning, how do other organisms achieve this? Scientists have discovered that many organisms possess the ability of "magnetic orientation."

It has been determined that the mechanism of bird navigation involves (1) sensing the Earth's magnetic field through sensors containing iron oxide nanoparticles, which transmit the magnetic signal to the brain, and (2) utilizing pairs of photochemically generated radicals in the retina, which exhibit quantum spin dynamics. This enables birds to detect changes in the Earth's magnetic field. Additionally, German scientist Sabine Begall discovered from satellite photos on "Google Earth" that cattle tend to align their bodies along the north–south axis while grazing. Biological quantum sensing mechanisms are far more sensitive than any current sensor in the world, which suggests that the electric sensors we use today still have great room for improvement.

8.2 Metrology and sensing

Metrology is the science of measurement, primarily focused on establishing a worldwide consensus on units. It serves as the foundation of technological advancements and also impacts human life. Fundamental physical quantities such as temperature, length, mass, and time, as well as derived physical quantities like velocity and acceleration, all fall within the research scope of metrology. Metrology is typically divided into (1) scientific metrology, which defines various units of measurement, (2) applied metrology, where measurements are applied in manufacturing and other societal contexts, and (3) legal metrology, which involves regulations and statutory standards for measuring instruments and measurement methods.

High-precision measurement is crucial to all branches of science. Quantum metrology is the study of achieving high-resolution and high-sensitivity measurements of physical parameters using quantum phenomena. It includes the development of measurement techniques that surpass the precision of classical measurement methods through the utilization of quantum entanglement and quantum measurements.

Sensors are devices used to detect various changes in the environment. Once a sensor detects a signal, there are typically several ways to process the information: (1) processing the acquired information locally, which is known as real-time edge computing, or (2) transmitting the acquired information to other electronic devices or cloud computing

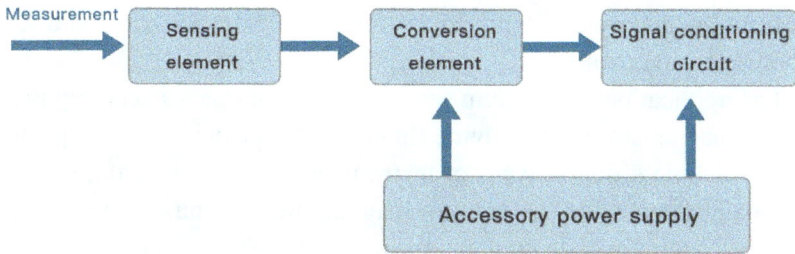

Figure 8.1. The sensor signals are transformed by the conversion components into processable information.

systems for centralized processing. Sensors are usually composed of sensing elements and conversion elements (Figure 8.1). Passive sensors, such as the typical forehead thermometer, fall into this category. However, sensors sometimes also include signal generators, transmitters, and receivers. These sensors are called active sensors. When the signal generated by an active sensor contacts the object to be measured, it reacts with the object. The reflected signal is then detected by the receiver and analyzed. Radar is an example of active sensors.

A quantum sensor, in simple terms, is a device that enhances human sensory perception — sight, hearing, smell, taste, touch, and proprioception — through quantum instruments to detect changes in the environment that were previously imperceptible. Quantum sensors leverage quantum entanglement to break through the limits of classical sensors, elevating the measurement precision of various fundamental physical quantities and derived physical quantities, such as temperature, magnetic fields, pressure, time, length, and weight, to the quantum limit. Quantum sensing employs quantum systems such as atoms, molecules, photons, and even artificial atoms to measure changes in the environment; each of these strategies may rely on distinct physical principles. The field of quantum sensing involves the design and engineering of quantum emitters and quantum detectors. Quantum sensing harnesses features such as quantum entanglement and quantum interference to break through the sensing limits of modern electronic sensors, making it easier for us to monitor various environmental changes. Quantum sensing and quantum metrology have wide-ranging applications in the aerospace industry, climate monitoring, construction, energy development, biomedicine, transportation, and water resource management.

The application of quantum sensors spans various aspects of physics, such as magnetism, light, gravity, time, and frequency. Current quantum sensors include different types of instruments, such as quantum gravimeters, quantum compasses, quantum magnetometers, and quantum clocks. Figure 8.2 lists some quantum measuring instruments, which we will cover in the next section. Quantum sensing technology is considered as important as quantum computation and quantum communication. It is generally believed that quantum sensing technology is less complex and smaller in scale compared to quantum computers. However, it is more closely related to everyday life needs, allowing the results of research and development to enter the commercial market more quickly. Currently, most quantum

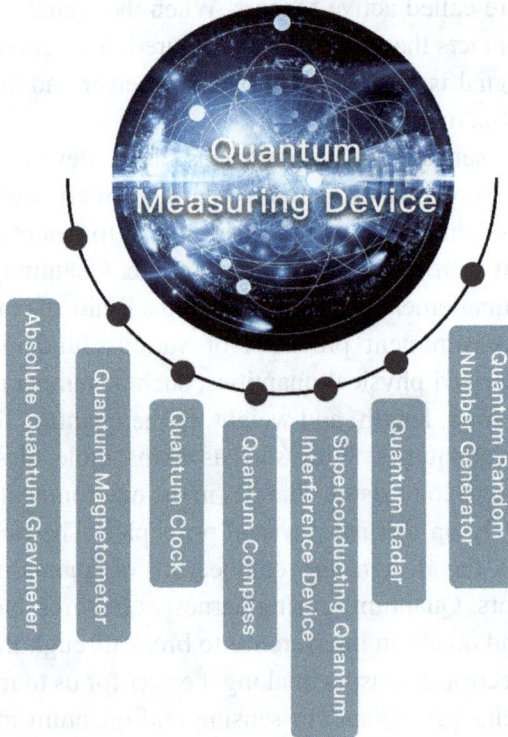

Figure 8.2. Various applications of quantum sensors.

sensing devices available in the market are still large-laboratory-level instruments. Shrinking the size of facilities and reducing the cost of quantum sensors in the short term still requires further effort from the research community. Quantum sensing technology has emerged as a frontier and key field in the new wave of technological revolution and industrial transformation. In 2016, CIQTEK was established, dedicated to the research, development, and industrialization of quantum precision measurement and high-end scientific instruments. Quantum precision measurement is a measurement technology that uses quantum technology to measure physical quantities with high resolution and high precision. CIQTEK commenced construction of the Quantum Science Instrument Valley in the Hefei High-tech Zone in China in May 2022. It aims to become the headquarters for the scientific instrument industry in China.

8.3 Applications of quantum sensors

8.3.1 *Quantum measurement*

In quantum physics, a measurement is a test or manipulation of a physical system, which yields a numerical result. The predictions given by quantum physics are probabilistic in nature. In the quantum world, knowing is equal to measuring, and you do not gain information without measuring. If you want to know, you need to choose the measurement method. The specific measurement method used will determine the measurement outcome, and the measurement will affect the quantum state being measured. These features of quantum measurement are closely related to Heisenberg's uncertainty principle, which states that momentum and position cannot be precisely measured simultaneously. The Heisenberg uncertainty principle was formulated based on experimental observations, and it implies a no-cloning property of quantum mechanics. In quantum mechanics, there exists a peculiar theorem called the "no-cloning theorem," which states that unknown quantum states cannot be cloned. That is to say, for an arbitrary quantum state, we cannot determine all of its quantum properties through measurement. Because if a quantum state could be cloned, one could first create many exact copies, and then use different measurement methods to precisely determine various quantum properties individually.

This would be against the Heisenberg uncertainty principle. The no-cloning theorem and the Heisenberg uncertainty principle are essentially two sides of the same coin; both measurement and cloning will affect the original quantum state.

Any quantum measurement requires a read probe to perform the measurement. A quantum probe projects a quantum state into one of several possible final states, and the measurement process is inherently uncertain and susceptible to external noise. In the field of quantum metrology, researchers have already developed the most precise measurement instruments to date. The atomic clock developed by the National Institute of Standards and Technology (NIST) in the United States is so accurate that it has an error of less than one second from the beginning of the universe until the present. A variety of novel quantum sensors have also emerged, capable of detecting submarines and stealth aircraft, as well as positioning, navigation, and timing (PNT). This "quantum PNT device" is an inertial navigation system that can realize navigation without relying on GPS, potentially changing the rules of underwater navigation and having an impact on national defence strategies.

(1) Absolute Quantum Gravimeter (AQG)

The AQG measures gravitational force through the quantum interference pattern produced by the free-falling of cold atoms. Before the advent of quantum gravimeters, gravity was measured using laser interferometry to determine the acceleration of reflectors free-falling in a vacuum. In AQGs, a group of cold atoms (usually rubidium) is released from a height, and then their vertical acceleration is measured through matter wave interference. Due to the utilization of quantum properties, the measurement accuracy can exceed one billionth of the gravitational acceleration of the Earth (9.80 m/s^2). Accurate gravity measurement can be helpful in many ways, such as global water resource management and natural disaster monitoring. Moreover, it can be used to inspect changes in roads, dams, and levee structures (Figure 8.3). Taking Earth scanning as an example, large-scale global monitoring can be achieved using satellites, combined with big data analysis of atmospheric and hydrological data. This approach is useful for predicting and monitoring various natural disasters such as floods,

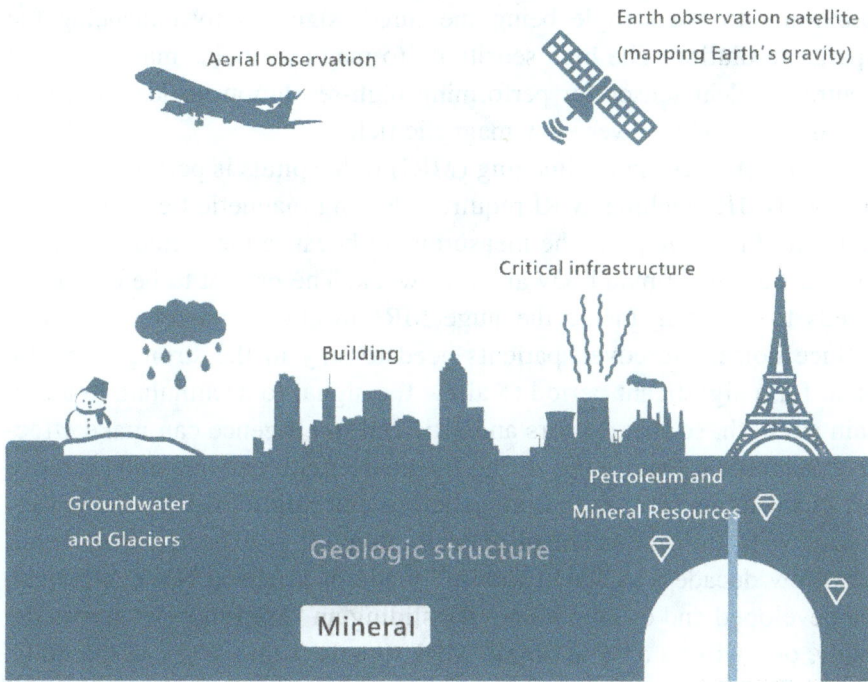

Figure 8.3.　Large-scale global monitoring using satellites and quantum gravimeters.

earthquakes, and volcanic eruptions. Quantum gravimeters can provide higher-resolution images compared to current satellite imagery.

(2) Quantum magnetometer: nitrogen-vacancy (NV) centre NMR
The spin of NV centres in diamond can serve as a sensitive nanoscale magnetic field sensor. Negatively charged NV centres are point defects in the diamond lattice with unique quantum spin properties that enable ultra-sensitive detection of magnetic force changes. An NV centre in diamond consists of a nitrogen atom and an adjacent carbon vacancy, and its electron spin state is 1. As mentioned in Chapter 5, NV centres can be used as qubits for quantum computing. In addition to serving as qubits, NV centres in diamond can also function as quantum sensors, as their spin is highly sensitive to changes in external magnetic fields. Since the size of the NV centre sensor is at the nanometre level, it can be placed in close

proximity to the sample being measured, significantly enhancing the spatial resolution. The high sensitivity to magnetic fields makes the NV centre an ideal sensor for performing high-resolution nuclear magnetic resonance (NMR) at very low magnetic fields.

Magnetic resonance imaging (MRI) in hospitals is performed using a large NMR machine. MRI requires a strong magnetic field of at least 2 Tesla (T) to complete the measurement because the signals from tissues inside the human body are very weak. The patient to be examined needs to be lying inside the huge MRI machine and remain still to reduce noise. Moreover, patients need to stay in the strong magnetic field for a significant period to allow the signal to accumulate to a certain strength, so that doctors and artificial intelligence can make effective diagnoses. Due to the strong magnetic field, patients must remove all metal items from their bodies before entering to avoid unnecessary harm. With the advancement of technology, it is expected that within next few decades, an "MRI mouse" made of artificial NV centres can be developed and used at home. By sliding an "MRI mouse" across the body, one will be able to obtain MRI signals from tissues in the body. These signals can then be sent to a computer to reconstruct overall images, which can be transmitted to doctors for diagnosis without the need for patients to personally spend much time staying in those massive machines at hospitals.

Cardiac arrhythmia is characterized by an irregular heart rate, which can be too fast or too slow. Magnetic induction tomography is an emerging quantum technology for diagnosing cardiac arrhythmias by measuring changes in the heart's magnetic field. It can be used to diagnose fibrillation and study its underlying mechanisms. The advent of quantum magnetometers is helpful in clinical applications such as imaging, patient monitoring, and surgical planning. Quantum magnetometers have advantages such as small size, light weight, low power consumption, and low cost, which can significantly reduce hospital operating expenses and the burden on patients.

(3) **Quantum clock**

An atomic clock is a clock that measures time based on atomic resonance frequency. It also serves as a standard for international time and is used to control signals of television broadcasts and GPS satellites. Germany's caesium atomic clock has an error of less than 1 second in 187 million years (uncertainty of 1.7×10^{-16}). Recently, a new type of quantum clock has been developed by the Massachusetts Institute of Technology (MIT), which achieved even more minor timing errors by measuring the behaviour of a group of atoms that are quantum entangled with each other. Measuring the behaviour of individual atoms is akin to flipping a coin. However, if a group of atoms becomes entangled, their oscillation frequencies become synchronized, resulting in lower errors compared to non-entangled groups of atoms. Quantum clocks are highly sensitive, allowing us to not only detect dark matter, gravitational waves, and other phenomena but also explore fundamental questions in physics, such as the following: "Does the speed of light change with the age of the universe? Does the cosmological constant change over time?"

(4) **Quantum compass**

The GPS is an essential function in modern technology products. It is used in mobile phones, drones, cameras, and vehicle navigation systems. It is hard to imagine life without the GPS now; it is probable that everyone will get lost inside a city. However, traditional GPS systems come with the risk of poor signal reception. Recently, scientists have developed a "quantum compass" that uses quantum technology for positioning. It primarily involves a "quantum accelerometer" that utilizes quantum phenomena to measure acceleration. In fact, there are accelerometers in mobile phones and laptops to measure changes in the object's velocity over time. However, conventional accelerometers have a drawback — they cannot maintain stable accuracy for extended periods without external satellites to calibrate location data. There are various current technologies that can interfere with GPS, posing a significant concern for maritime operations and national security. Quantum compasses, because they operate independently of satellites, offer precise positioning at any location on Earth and are immune to external interference. However, current quantum accelerometers are still in the experimental phase, with large volumes and high costs. They also need to operate in an extremely low-temperature environment. In the future, the major effort will be emphasized to reduce both the size and the cost of instruments and to commercialize. Since quantum accelerometers are not affected by external influences, in the future, this technology can be used in places where GPS signals may not be received due to environmental factors, including dead spots inside shopping malls and deep-sea nuclear submarines.

The greatest strength of a submarine is that it is hard to detect after it is submerged. Good stealth makes submarines an important intimidation weapon in wartime attacks and peacetime, because submarines can be sent to areas of military instability as a show of force rather than direct provocation. However, the drawback is that once the submarine enters the sea, it will be isolated from GPS signals and even stars, so deep-sea guidance and positioning become difficult. Although modern submarines use gyrocompasses and inertial guidance systems to make guidance corrections in the deep sea, the effects are limited. The errors in the correction guidance system will accumulate over time, and the submarine will no longer be able to identify its position and direction. Therefore, the submarine must occasionally resurface and use GPS to recalibrate its navigation and positioning, making it easily detectable by the enemy. The AUKUS partnership (AU, UK, and US, a trilateral security pact between Australia, the United Kingdom, and the United States) exploits properties such as quantum entanglement, quantum interference, and quantum state squeezing to conduct revolutionary quantum submarine deep-sea navigation. According to quantum start-up Q-CTRL in Australia, in the future, submarines will only have a maximum error of one mile (1.6 km) within 1,000 h of underwater operation. As this quantum compass technology is being developed under the AUKUS treaty, in addition to being used in Australia, it will also be shared with the United States and the United Kingdom. The Chinese navy is also looking to develop a quantum compass for its submarines that would enable them to navigate without the help of the Beidou satellite navigation system (BDS). US defence technology giant Lockheed Martin claimed to use NV diamond to make a quantum compass to improve the navigation capabilities of the US navy. When illuminated by a laser, the intensity of the light the compass emits changes according to the surrounding magnetic field. Through the Earth's magnetic field, this change in light can improve navigation capabilities, especially in extremely remote areas. In October 2021, a US nuclear submarine hit an unknown underwater object in the deep waters of the international regions of South China Sea, with no knowledge of what it collided with and why the accident occurred. With the presence of quantum compasses in the future, such deep-sea accidents will be greatly reduced. Militaries around the world are interested in this new technology because of the limitations of the widely used space-based navigation systems nowadays.

(5) Superconducting Quantum Interference Device (SQUID)

SQUID and Magnetic Anomaly Detector (MAD) both detect submarines by detecting magnetic fields, but the sensitivity of SQUID is much higher than that of MAD. The Josephson junction within SQUID is also an essential component in superconducting qubits. SQUID is an ultrasensitive magnetometer that can measure magnetic fields as low as 5×10^{-18} Tesla (T). For comparison, the magnetic field of a refrigerator is approximately 0.01 T, the Earth's magnetic field typically ranges around 5×10^{-5} T, and biological magnetic fields are between 10^{-6} and 10^{-9} T. Currently active anti-submarine aircraft are all equipped with MAD to detect submarines by monitoring abnormal changes in Earth's magnetic field, but its detection range is limited to 500 m. In 2017, China published an article about installing SQUID on a helicopter and successfully detecting iron-containing materials deep underground. Based on this information, the United States estimated that this technology can detect submarines from a distance of 6,000 m. Therefore, the United States is concerned that China will easily locate its submarines in the deep sea. On March 31, 2021, an article in the US magazine "National Interest" claimed that China's advancements in SQUID technology would render nuclear submarines useless. The maximum operating depth of the US Seawolf-class or Virginia-class nuclear submarine and the Russian Akula-class nuclear submarine is greater than 500 m, making them capable of evading traditional magnetic anomaly detectors (MADs), but not the SQUID. The average depth of the Earth's oceans is only about 3,600 m, and the deepest part of the South China Sea is only 5,559 m. Therefore, once China has detection technology capable of reaching depths of over 6,000 m, nuclear submarines will no longer be able to evade detection.

(6) **Quantum radar**

Radar is an abbreviation for radio detection and distance measurement. Radar systems emit electromagnetic waves in a directional manner into space and then receive the reflected radio waves from objects in the sky. This allows for the calculation of the object's direction, altitude, velocity, and shape. With the rapid development of stealth technology and electronic jamming technology, stealth aircraft have become the primary offensive force on the battlefield, rendering classical radar useless. Moreover, if there were stealth missiles in the future, it would be impossible to guard against sudden attacks. Consequently, effective detection of stealth aircraft and projectile weapons has become a crucial battleground in military affairs.

In recent years, the United States has actively developed stealth aircraft, stealth missiles, and stealth drones. Stealth technologies mainly achieve invisibility by airframe design and stealth coatings. These measures reduce radar's ability to detect aircraft, allowing stealth aircraft to hide in the vast skies. As shown in Figure 8.4, classical radar systems emit

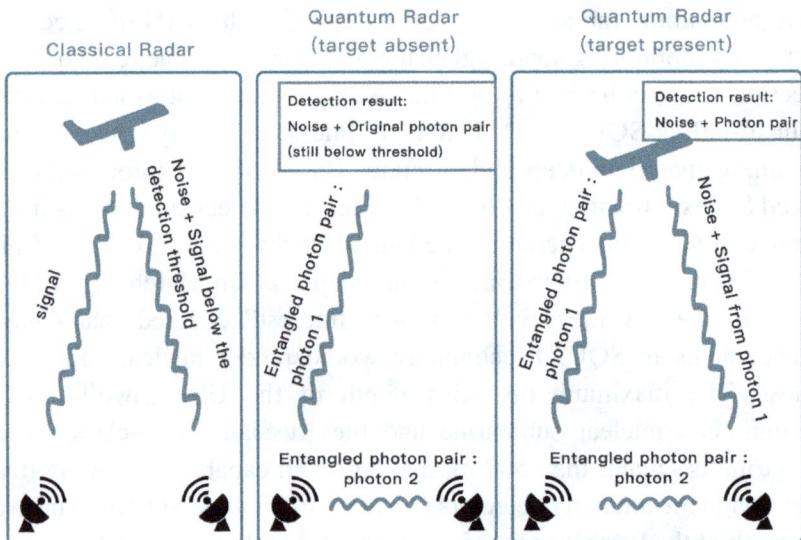

Figure 8.4. Schematic Illustration of the working principle of quantum radar and classical radar. On the left, classical radar emits electromagnetic waves that encounter flying objects and then reflect back to the receiving antenna. On the right, quantum radar emits a pair of entangled photons, significantly enhancing sensitivity and resistance to interference through quantum entanglement.

radio waves from the transmitting antenna. When these radio waves encounter flying objects, the receiving antenna will receive the reflected waves, allowing detection of the presence of an aircraft in the airspace. However, due to the special coating and design shape of stealth aircraft, classical radars not only fail to detect signals due to the reduction of reflection but may also encounter interference from false signals and give false alarms, losing their air defence functions.

Quantum radar is the terminator of stealth technologies. Usually, there are three major types of quantum radars:

Type 1: Quantum Transmitter and Classical Receiver: The quantum sensor transmits unentangled quantum states of light. The signal transmitted comes from a quantum source to produce a single-photon source.

Type 2: Classical Transmitter and Quantum Receiver: The quantum sensor transmits classical states of light but uses quantum photo-sensors to boost its performance. The system usually has a single-photon detector to detect the signal has quantum components.

Type 3: Quantum Transmitter and Quantum Receiver: The quantum sensor transmits quantum signal states of light that are entangled with quantum ancilla states of light kept at the transmitter. Quantum transmitters and receivers have advantages when used together. An example is the joint detection of a pair of entangled photons at the receiver. This type of quantum radar is based on quantum illumination protocols and uses entangled photon pairs to enhance target detection capabilities and provide strong immunity to environmental noise.

Quantum radar is very suitable for various military applications and is the focus of military competition among various countries. As shown in Figure 8.4, the transmitting antenna emits a pair of entangled photons, with one directed into the sky and the other sent directly to the receiving antenna. Due to the entanglement between the photons, when the quantum wave encounters a stealth aircraft in the sky, its entangled properties enable easy detection of the object. The advantage of quantum radar is its resistance to interference. As quantum measurements will alter the characteristics of photons, regular checks on photon properties can reveal whether they have been disturbed in the airspace, effectively preventing deception. With the gradual maturity of quantum radar in the future, stealth technologies can be monitored and tracked. Quantum radar is the nemesis of expensive stealth aircraft.

Theoretically, quantum radar possesses the capability of detecting stealthy technologies, potentially serving as the "all-seeing eye" (i.e., clairvoyance) on future battlefields. However, due to various physical limitations, the feasibility of quantum radar is still questioned by many scientists. Nevertheless, the US Department of Defense frequently cites media reports claiming that China has developed a new type of quantum radar capable of detecting stealth aircraft from up to 100 km. Therefore, in 2021, the US Senate Committee passed the "Endless Frontier Act," requesting the government to allocate over $110 billion for basic and advanced technology research, with a focus on increasing investments in quantum research. This bill aims to ensure that quantum technologies can be applied to national defence in the future, thereby maintaining America's global advantage. Interestingly, the US Department of Defense mentioned in its official report on quantum technology in 2021 that quantum radar may not have significant military utility, and its true defence applications are still a subject of debate. Since the quantum radar is still in the primary stages of research due to the lack of effective entangled microwave sources, some possible civilian applications of an alternative version of quantum radar have also been proposed. This involves, not the use of microwave as in radar, but the use of light as in lidar. Quantum lidar has been proposed for use in monitoring various environmental changes, such as the presence of toxic gases. UK-based company QLM said its detector provides a panoramic multi-directional view of equipment and associated gas emissions from up to 200 m away. There are also other new ideas for lidar using quantum detection. In atmospheric detection, China uses infrared light quantum lidar during the day to detect the movement of gases in the atmosphere, thereby accurately predicting the weather.

(7) **Quantum Random Number Generator (QRNG)**

When playing Monopoly as a child, we all knew that the dice is a crucial tool; the numbers rolled can determine the outcome of the game. Almost all board games require a dice to introduce randomness. The most common dice is the six-sided one, but there is also a polyhedral dice (i.e.,

Platonic solid) that generates a wider range of random numbers. Dice are ancient random number generators.

Random number generators are used to provide randomness for online video games, cryptography applications, and other fields. These generators typically employ algorithms to produce a sequence of random numbers. Random number generators yield random numbers whenever needed, and are like virtual dice in a programme. Physical objects such as dice, which can generate sequences of random numbers, are generally referred to as True Random Number Generators (TRNGs), while algorithms that generate random numbers are called Pseudo Random Number Generators (PRNGs). Sequences generated by PRNG are not truly random; they are determined by an initial seed value, which completely decides how the sequence of numbers is generated. Therefore, if providing the same initial seed value, all numbers in the sequence will be reproduced, unlike TRNGs, which are completely unpredictable. Random number generators are now an essential component in computer simulations and cryptography, making a perfect random number generator a necessity.

A TRNG typically produces a very limited number of random numbers per second, much slower than a PRNG. To improve the efficiency of random number generation, a TRNG is often used as the initial "seed" value for a PRNG, allowing faster generation of realistic and non-repeating sequences of random numbers. Strictly speaking, throwing dice still falls within the scope of application of classical mechanics. As long as the initial conditions and applied forces are properly controlled, theoretically, the number produced by dice is predictable. Therefore, some argue that a dice is not a TRNG, but rather a hardware random number generator (HRNG). Encryption, computer simulations, online games, and more modern technology rely on fast and fair dice, so the pursuit of a perfect random dice is an ongoing and important field of research.

Einstein's challenge to the Copenhagen school's interpretation, "God does not play dice," underlines the probabilistic nature of quantum theory. QRNGs exploit the inherent probabilistic nature of quantum physics to create a truly perfect dice. Since the source of randomness in a QRNG is well understood, quantum components that generate random numbers based on quantum processes have been used in information encryption.

In 2020, ID Quantique (IDQ) announced the integration of QRNG chips into Samsung smartphones to protect users' sensitive data. The current QRNG research direction primarily focuses on creating more economical, faster, and smaller quantum chips for random number generation.

8.4 Concluding remarks

Human beings have limited abilities and sensitivities, and what the eyes, ears, nose, tongue, body, and mind can perceive is never the whole world. Throughout history, many people have wished to understand the universe more deeply. However, due to the lack of sensitive enough sensors that go beyond the perception of our senses, various myths and legends related to perception have been created. Among them, many fantasies are "pseudo-science" disguised with scientific terminologies, but some of them do have certain levels of scientific validity. Science, however, is just one part of knowledge and has its specific research scope and methodologies. The progress of human knowledge and the growth of technology in the past are largely due to the correct use of scientific methodology, which has expanded the boundaries between known knowledge and the unknown. As shown in Figure 8.5, knowledge can be roughly categorized into the following:

(1) **Application — the known knowns**: applied sciences such as engineering, where the fundamental principle has been well understood. A significant amount of manpower and resources are needed to explore applications. Quantum technology has just begun to enter this domain.

(2) **Exploration — the known unknowns**: the knowledge to be understood, which resides at the intersection between what is already known and what remains unknown. In this domain, a substantial body of evidence and data are already available, and they align with existing scientific knowledge. This area is the primary focus of scientific research. Interestingly, the smaller the known territory, the smaller the area that needs to be explored. This explains why ancient scientists could span across multiple fields. In modern times, due to the vast extent of established knowledge, the boundary between the known and the unknown

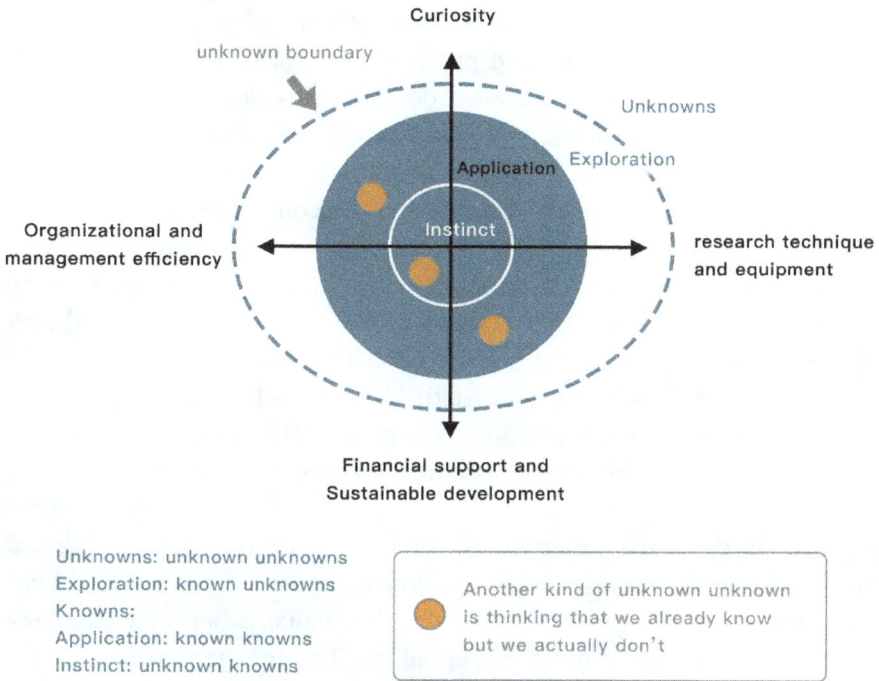

Figure 8.5. Categorization of Knowledge: (1) Application: Known Knowns. (2) Exploration: Known Unknowns. (3) Instinct: Unknown Knowns. (4) Unknown: Unknown Unknowns. The study and expansion of knowledge, as depicted in the figure, require various driving forces to propel them forward. Human curiosity, research techniques and equipment, organizational and management efficiency, and financial support are all indispensable in this endeavour.

that needs exploration has become extremely large. Therefore, more manpower and material resources are required to continue to expand the known domain of mankind to the unknown territories, and this is the spirit of the continuous growth of scientific exploration.

(3) **Instinct — the unknown knowns**: This domain mainly consists of rules of thumb. This domain is supported by a lot of observational data and is often reproducible. For example, historically, astronomy began in this domain and gradually evolved into a scientific discipline. Modern sociology, economics, and management studies share similarities with this category; they exhibit a certain degree of reproducibility but are not as precise and controllable as physics.

(4) **Unknown — the unknown unknowns**: a completely unknown area,
    where even phenomena and problems are not understood, lying far
    beyond the edge of the known domain. This domain may even be
    outside the scope of the four quadrants. Myths and legends fall into
    this category. There is insufficient empirical data in this field, and the
    replicability is extremely low, making it incompatible with the scien-
    tific methodology. We can only hope that more replicable evidence
    will be discovered in the future. Due to the high level of uncertainty
    and irreproducibility, this field can usually only serve as a casual con-
    versation topic. It is not suitable as a short-term goal for scientific
    exploration. Confucius once said, "You do not yet know about the
    living, how can you know about the dead?" This quote advises every-
    one to avoid exploring the phenomena in this domain.

In fact, there is another category of "special misconception" spread
across all domain, represented by the dots in Figure 8.5. Some individuals
believe they already possess knowledge, but actually their understanding
is partial or even based on incorrect beliefs. The emergence of such misconceptions can sometimes be attributed to mere ignorance due to insufficient understanding, but at times, it is indeed a result of deliberate
falsifications driven by selfish desires. These misconceptions often lead to
adverse effects to society and sciences. On the way to creating the future
of the "Second Quantum Revolution, we must avoid going astray, lest we
waste time and energy and achieve nothing.

Quantum sensor technology has now gradually entered daily applications (known knowns). One still requires a large number of quantum
engineers and much resource investment to develop various devices that
break through the limits of classical sensors. The future quantum sensing
technology will enable human senses to be further extended, helping to
fulfil long-held human dreams of going beyond the six senses.

Chapter 9

Quantum Communication and Quantum Internet

Quantum computing is like a spear, quantum communication is like a shield, and quantum sensors are like super six senses. The integration of all quantum technologies is the quantum Internet of Things.

— Ching-Ray Chang (1957–)

The speed of communications is wondrous to behold. It is also true that speed can multiply the distribution of information that we know to be untrue.

— United States, Edward R. Murrow (1908–1965)

9.1 The importance of encrypted transmission

There are two main technological process involved in modern network communication: One consists of establishing a transmission network, which must be able to effectively and quickly send information from the sender to the receiver, and the other consists of encrypting information so that it will not be intercepted by a malicious third party during transmission. When using the Internet, you will notice that all web addresses have two types of beginnings: HTTP and HTTPS. Both network systems can browse different web pages. The difference is in encryption. The extra "S" in HTTPS stands for "Security." HTTPS adds an additional layer of security through a "secure protocol," using encryption to solve privacy issues that cannot be protected in HTTP. Many websites still use HTTP, so it is essential to be cautious with personal data when using these websites, as HTTP offers little resistance against hackers.

There are three main challenges found in modern network communication, which are similar to the problems encountered in the transmission of confidential information in ancient times: (1) a safe, effective, and fast delivery network, (2) safe and reliable packaging containers for items to be delivered, and (3) confidentiality of the transmitted content itself. The tools employed to address these three challenges have evolved over time due to technological advancements, but the fundamental principles of communication security and the concepts remain similar to those used in ancient times (Figure 9.1). The ancient Chinese posthouse transmission system is similar to the modern network information transmission system. In the past, letters were sent by people riding fast horses, but now they are transmitted on the network through a digital ciphertext. The posthouse was a place where people who delivered documents and military information could rest and change their exhausted horses. China was one of the first countries in the world to have an organized information transmission system. The posthouse system existed in China for more than 3,000 years. With the information transmission network, it is necessary to encrypt the reliable container that is used for transmitting the information to ensure the security of the information in the transmission process. The earliest encryption methods used in ancient China were sealing with mud, tying bamboo slips with ropes, and stamping the seals on the mud at the knots. Since the mud and the seal stamp are unique, it was not possible to restore the sealed mud and seal stamp after unpacking. Therefore, it was guaranteed that no one could steal information during the transmission, just like today's quantum communication system with entangled photon pairs. In ancient times, important information was even double encrypted, with officials of different ranks sealing information twice to prevent it from falling into the hands of unauthorized people. There was also a code for secret military communication in the Northern Song Dynasty. "Wu Jing Zong Yao" (武經總要) records that various common war terms, such as "besieged," "needing reinforcements," and "sending weapons," were summarized into 40 short messages using a method of shorthand characters.

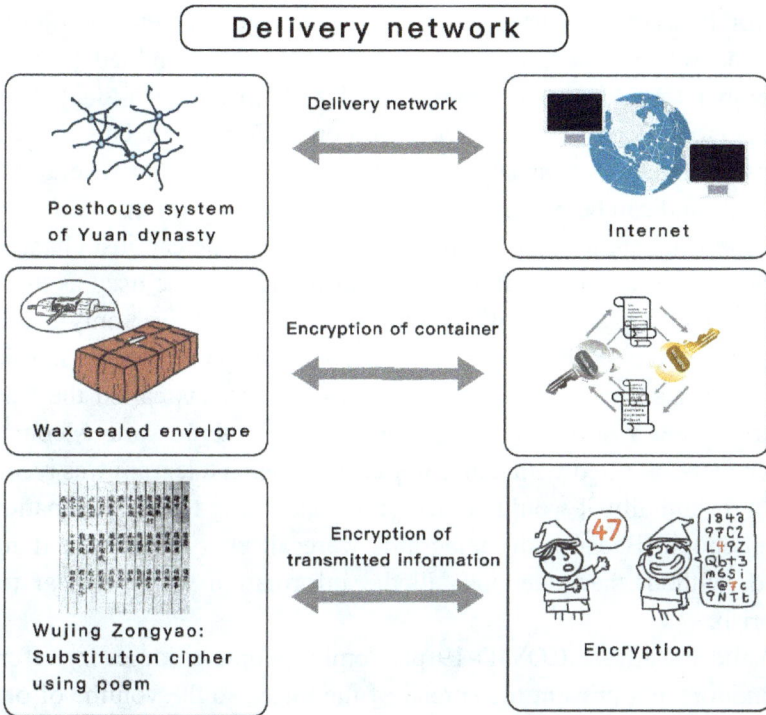

Figure 9.1. The concept of information encryption has not changed much from ancient times to the present. It is just a trilogy of (1) encryption of the transmission network, (2) encryption of the containers used for transmission, and (3) encryption of the transmission information. However, as technology has advanced, the tools used have changed. Generally speaking, the more complex and numerous the encryption steps and methods are, the harder it is to crack, but also the more difficult it is to use.

Before launching a military campaign, the command headquarter would designate a 40-word poem as the key. Each word in the poem corresponded to a text message, and poems would be used for encrypted communication between the front and rear of the battlefield.

Another simple way to encrypt message content is to scramble the text so that even if the message is intercepted, the content cannot be understood. But, how can information be scrambled? Ancient Chinese encryption used a method similar to solving lantern riddles during the Lantern Festival. In the era of Wu Zetian (武則天), the ministers rebelled, and the emperor intercepted a letter with only two words [green goose](青鹅). Everyone was puzzled, but Wu Zetian easily disassembled [green] as [December (十二月)] and [goose] as [I and myself (我自与)], that is, I will personally participate in the rebellion in December. This is because Chinese characters are usually formed by a combination of several different words and can be broken down into their original meanings. If a message is in English, it is more straightforward and one can just change the order of the letters; for example, the fourth letter can be used as the first letter, that is, replace A with D, and so on. The receiver only needs to restore the order of the letters. This oldest encryption method is said to have been used by Julius Caesar in wars, hence it is called the Caesar cipher. During World War II, high-ranking generals in various countries had a codebook filled with dense numbers. When a telegram was received from a distant ally, it would be decrypted according to the information in the codebook. Because the telegrams were all encrypted, even if intercepted, without the codebook, all the information would appear to be gibberish.

Amid the severe COVID-19 pandemic, people made every effort to stay indoors and prevent the spread of the virus, so the volume of online shopping increased significantly. Due to the maturity of modern information technology and the rapid development of network technology, we can conveniently purchase all kinds of necessary items through the Internet. The process typically involves logging into an online shopping platform, entering personal account information and passwords, adding the goods into the shopping cart, filling in the delivery address and contact number, proceeding to checkout, and inputting credit card information, after which the transaction is complete. We can relax at home and wait for the logistics industry to deliver the goods to your door. It is hard to imagine how pandemic isolation and lockdowns would be carried out without online shopping. However, in the process of online shopping, all personal information is transmitted over the open network and can be stolen by malicious hackers. Ensuring that data remain secure and unaltered during transmission is thus a critical concern in this digital age. Without reliable encryption protection, today's massive financial transactions cannot be carried out on the Internet.

For regular online shopping, e-commerce companies will use encryption to protect customer information. The current advanced encryption methods can be divided into symmetric encryption and public-key (asymmetric) encryption. If the same key is used for encryption and decryption, it is convenient to use and is called symmetric cryptography. If different keys are used for encryption and decryption, it is called asymmetric encryption. The asymmetric encryption algorithm is more complicated and involves advanced mathematical principles. In encrypted communication, the sending and receiving parties need to know the encryption method and key in advance. Otherwise, the receiver will not be able to decrypt the information after receiving it.

With the encryption protocol, effectively and safely exchanging keys on the network is a critical issue, especially in communication networks where absolute security cannot be guaranteed. Asymmetric encryption, currently the common choice in the digital world, can address the key distribution problem and thus has become standard procedure. As shown in Figure 9.2, in asymmetric encryption, two kinds of keys are generated each time: a public key and a private key. The concept is very simple: Imagine a special lock with three positions, A (locked) — B (open) — C (locked), with the rotations of the two keys being in opposite directions. If the rotation order of the public key is clockwise, A->B->C, then the rotation order of the private key is counter-clockwise, C->B->A. When Bob wants to send a message, he prepares a private key and a public key

Figure 9.2.   The special lock has a public key and a private key. The public key and the private key rotate in different directions when opening the lock.

and sends the public key and a box with a special lock to Alice. The lock of the box is in the open state of B. After Alice receives it, she just needs to put the information in the box and then use the public key to lock the box to state C and send it back. When Bob receives it, he uses the private key to unlock and open the box. The clever part of this idea is that even if Eve, a malicious eavesdropper, intercepts the public key during transmission and captures the box sent back by Alice, she will not be able to open the box because the public key cannot be rotated from C to B. Only Bob's private key can open the box, so the transmitted information is safe. RSA encryption is such an asymmetric encryption. There is a mathematical relation between the public key and the private key in asymmetric encryption, but calculating the private key from the public key requires enormous computation, which is almost impossible with today's classical computers.

There is also a need for identity verification. How does Bob determine that the information in the box is from Alice? This encryption system completely solves this authentication problem. The method is as follows: Alice also prepares a small locked box, puts the message into the small box, locks it in position A with her own private key, then puts it into Bob's box and locks it in position C with Bob's public key. Alice then sends the box back to Bob. When Bob receives it, he uses his private key to open his special box, and then uses Alice's public key to open the small box. If he can unlock the small box, it can be confirmed that the information comes from Alice. This method is a bit like the earlier mentioned ancient practice of double sealing. Encrypting with a public key and decrypting with a private key form an encryption method. Conversely, encrypting with a private key and decrypting with a public key can be done authentication, which we now refer to as a digital signature. This signature can ensure that the information is indeed sent by the holder of the private key, or the transaction is indeed initiated by that person.

Of course, there is no actual box being transported over the Internet. The current online virtual box of the encryption system completely relies on the mathematical complexity of generating public and private keys. Difficult mathematical problems need to be solved to produce the private key from the public key. Therefore, even if hackers intercept the information encrypted with the public key, they still cannot decrypt it without the private key. The harder the mathematical puzzle in an encryption system, the more secure the transmission of information. Among the current asymmetric encryption algorithms, the famous ones are the RSA encryption algorithm and the elliptic curve cryptography. They both rely on mathematical problems to ensure security. Therefore, some say that information confidentiality is essentially rooted in trusting mathematics and it is indeed "In Math We Trust."

The widely used RSA encryption algorithm in networks operates on the principle of factorizing a large integer. After multiplying two large prime numbers, it is extremely difficult to decompose them back to the original two prime numbers. The inherent asymmetry in prime factorization guarantees the reliability of encrypted transmission. It takes an extraordinarily long time for a classical computer to factor a large number, so the larger the number, the more reliable the security. Although today's classical computers cannot solve these mathematical problems, quantum computers can handle them. The answers to this kind of mathematical problem can often be transformed into a periodic structure. Once the period is found, the problem is solved. Due to the superposition property of quantum computing, which enables simultaneous computation of numerous states, and the use of quantum Fourier transform to speed up the search for the function's period, quantum computers can find solutions faster than classical computers. The development of quantum computers has been progressing rapidly in recent years. Y2Q will arrive sooner or later. Various encryption systems that protect our privacy have encountered powerful threats and are no longer safe. In order to combat such threats, large technology companies such as Google and Cloudflare from the National Institute of Standards and Technology (NIST) have actively responded and taken action to provide post-quantum encryption technology resistant to quantum computing.

According to the estimation of experts, quantum computers will be able to decrypt almost all traditionally encrypted data after 2030. It has become a top priority to find countermeasures as soon as possible to ensure the absolute security of future information transmission. In addition, many enterprise organizations are very concerned about the "harvest now, decrypt later (HNDL)" attack method. Those who are interested in the secret only need to collect the encrypted messages now, and they can decrypt them when quantum computers with powerful computing power mature later. Since email transmissions often contain a large amount of sensitive confidential information, if post-quantum cryptography (PQC) is not used immediately to protect it, it actually means that secrets will be leaked in the future, and the consequences will be disastrous. Inventing better ways to protect the security and privacy of information is not a problem for the future era of quantum computers, but a problem that needs to be considered now.

9.2 Post-quantum cryptography

The era of PQC has begun since the quantum computer is coming. In classical cryptography, the emerging trend in PQC involves creating encryption scheme that can resist quantum properties in response to the advent of universal quantum computers. In another new direction, quantum cryptography (QC) based on quantum mechanics has been developed to ensure information security by perceiving eavesdropping so that eavesdroppers cannot steal information. However, PQC is a pure classical method, while QC is really based on quantum phenomena. The most ironic point is that PQC and QC have nothing to do with each other. PQC has always been a confusing terminology for laymen; why does PQC (post-QC) come before QC? A widespread joke in the cryptography community is "why do people want to study PQC even when QC does not exist yet." Another similar one is that post-quantum means no quantum.

Cryptosystems can be divided into two categories: symmetric cryptography and public-key (asymmetric) cryptography. Symmetric encryption systems are often used in mobile phone hardware encryption, network HTTPS, and are used by many ATM cards, debit cards, and banking systems. There are two main types of public-key cryptographies. One is elliptic-curve cryptography (ECC), which is used as a digital signature mechanism in Bitcoin. The other is the natural person certificate for the government, which uses RSA encryption to confirm user identity. Symmetric cryptography is like a safe. The same key must be used regardless of whether it is opened or closed. Public-key cryptography is like a personal mailbox at the post office. Anyone can put things in the mailbox, but only the owner has the key.

9.3 Quantum cryptography

As mentioned in the previous section, QC has nothing to do with PQC. One uses quantum effect and the other belongs to classical cryptography. Quantum entanglement can be used for information encryption, but it cannot enable superluminal information transfer. This is because transmitting information requires encoding specific information on one of the particles in an entangled pair, which destroys quantum entanglement. Quantum communication technology does not violate any scientific principles. It does not transmit information at superluminal speed, but is a safer encryption technology.

Both classical communication and quantum encrypted communication use photons to transmit information. However, in classical communication, each signal consists of a large number of photons, so their quantum properties are lost, making it impossible to utilize quantum characteristics for eavesdropping detection. On the other hand, if the digital code is encoded on the quantum state of photons, according to the quantum no-cloning theorem, the quantum state of a photon cannot be perfectly cloned. If someone attempts to steal and read the information, their action will alter the quantum state, leading to its detection. If Alice and Bob send photon states with quantum properties between each other, Eve, the eavesdropper, cannot steal the photon information without Alice and Bob knowing about it. If there is no eavesdropping, the correct information can be obtained using the method agreed upon by both parties. The security of classical communication relies on the fact that the eavesdropper has limited computing power, making it infeasible for them to break encryption keys. However, universal quantum computers have enough computing power to crack RSA and other ciphers. Although the new PQC can resist the deciphering of Shor's algorithm, it has not been proved that these PQC algorithms cannot be deciphered under the universal quantum computer. Using quantum encryption communication will provide absolute security for key distribution and one-time encryption will ensure absolute security for information transmission. Quantum Key Distribution (QKD) is protected by the laws of physics in the transmission channel. Quantum communication, which combines the characteristics of quantum mechanics and information theory, offers exceptionally high communication confidentiality.

QKD achieves quantum protection of key distribution by perceiving eavesdropping, making it impossible for eavesdroppers to obtain any transmitted content. Because the quantum properties ensure that the stolen key will be detected, new QKD can be distributed separately to ensure absolute security. Therefore, eavesdroppers will not be able to steal information through QKD deciphering. Traditional encryption mainly relies on the complexity and difficulty of mathematics, while quantum encryption borrows the characteristics of quantum mechanics, which can theoretically ensure the security of quantum communication. These quantum properties are as follows:

(1) **Quantum superposition:** A truly random key can be generated using the probabilistic nature of the quantum world, enabling one-time pad communication.
(2) **No-cloning theorem:** Any attempt to read a quantum state results in a change in that state, making it impossible to clone unknown quantum states and rendering any eavesdropping on a quantum network unfeasible.

The concept of quantum encryption was first introduced by Columbia University graduate student Stephen Wiesner with his idea of "quantum money." In the early 1970s, counterfeit banknotes were prevalent in the United States. Wiesner suggested using the principles of quantum mechanics to create a digital quantum currency that cannot be counterfeited, which utilizes the polarization state of photons as the password for banknotes. To make counterfeit banknotes, one would need to know the password of quantum banknotes, which requires measuring photons. However, measuring will destroy the quantum password, thus ensuring the anti-counterfeiting purpose of quantum currency. Unfortunately, at that time, the cost of quantum currency was too expensive, and market acceptance was unknown, so no one was interested in producing it. It was Wiesner who first proposed several of the most important ideas in quantum information theory, including quantum currency, QKD, and superdense coding. Wiesner once submitted his paper "Conjugate Coding" to *IEEE Transactions on Information Theory*, but it was rejected. Many of his studies were not published officially, but were disseminated through manuscripts, which significantly impacted the field of quantum information. In 2006, Wiesner shared the Rank Prize in Optoelectronics with Charles H. Bennett and Gilles Brassard for QC. In 2019, Wiesner received the Micius Quantum Prize, along with Bennett, Brassard, Artur Ekert, Anton Zeilinger, and Pan Jianwei, for quantum communication. During the rebellious era of the 1960s in the United States, all forms of authority were questioned. Young rebels — known as hippies — became trendy and so did Wiesner, as a quasi-hippie refugee from the 60s. Wiesner's father was the president of the Massachusetts Institute of Technology from 1971 to 1980, and there was a huge contrasting father and son image between the two. Wiesner immigrated to Israel later. He worked as a construction worker. He passed away in Jerusalem in 2021.

In 1979, at the beach of a posh hotel in San Juan, Puerto Rico, Wiesner's college classmate Charles Bennett mentioned Wiesner's forward-looking idea to the cryptographer Gilles Brassard during a chat, and the ideas resulted in the first paper ever published on quantum cryptography in 1982. Later, they published the first QC protocol based on the uncertainty principle and the no-cloning theorem in 1984, which was later called the BB84 protocol. This protocol marked the beginning of QKD research. Since then, the use of quantum properties to make ciphers has attracted attention and the field has developed rapidly. Artur Ekert of Oxford University subsequently proposed an entangled version of the QKD protocol in 1991, which involved using the distribution and measurement of entangled photons to share keys. This protocol became known as the E91 protocol. After the E91 protocol was proposed, Bennett, Brassard, and N. David Mermin proposed the entangled version of the BB84 protocol, called the BBM92 protocol. The critical difference is that BBM92 uses only two states instead of the four states used in the E91 protocol and BB84. Later, Bennett also proposed B92, which is a QKD method that uses entanglement distillation protocols to prepare and transmit non-orthogonal quantum states. Now, many QKD protocols are defined for different application purposes, and these QKD protocols are used internationally and are also an important basis for quantum communication.

The QKD protocol uses the quantum properties of photons to encode and transmit keys in binary. Due to the no-cloning theorem, any eavesdropping by Eve will interfere with the quantum states, allowing both the sender and receiver to immediately stop the communication and change the content of the communication before the information is stolen. Quantum encryption is like writing information on a soap bubble. If anyone touches it during transmission, the soap bubble will burst. Regardless of the specific QKD protocol, any eavesdropping attempt is detected by both the sender and receiver, which is why QKD is superior to traditional key distribution. In the following, BB84 and E91 protocols are briefly explained:

(1) **BB84 protocol:** When photons are emitted from the laser, a superposition state with random polarization directions is formed. Alice randomly chooses a combination of bases (which can be a rectilinear basis "+" or a diagonal basis "X") to generate qubits (0 or 1) for encoding and transmitting information. When Bob receives the information, he does not need to know in advance which basis Alice uses, so he also randomly selects the basis (+ or X) to measure each received quantum state. Bob records each selected basis and the respective photon measurement results, and reports to Alice through the classical channel. Alice keeps the basis and quantum measurement results that are identical for both parties, using them as the key for communication. As shown in Figure 9.3(a), the sender Alice and the recipient Bob exchange quantum information through optical fibre, and use a parallel classical channel to communicate measurement results. If Eve attempts to eavesdrop, she must measure the photons to obtain photon polarization. However, measuring in different bases than those chosen by Alice will alter the photon's polarization state, resulting in errors when comparing the keys. Alice and Bob can compare a portion of the keys to check whether someone is eavesdropping based on the error rate of the measurement data. If the error rate is higher than the statistical error, the original key must be discarded, and a new key is established using QKD protocol. When the quantum key has sufficient length, the transmission of information will be 100% secure and impossible to eavesdrop.

(2) **E91 protocol:** In the E91 protocol, entanglement is used to ensure communication security. Information is transmitted using a set of entangled photon pairs created by either Alice or Bob. Alice and Bob obtain one of the entangled photons. Since the two photons in the

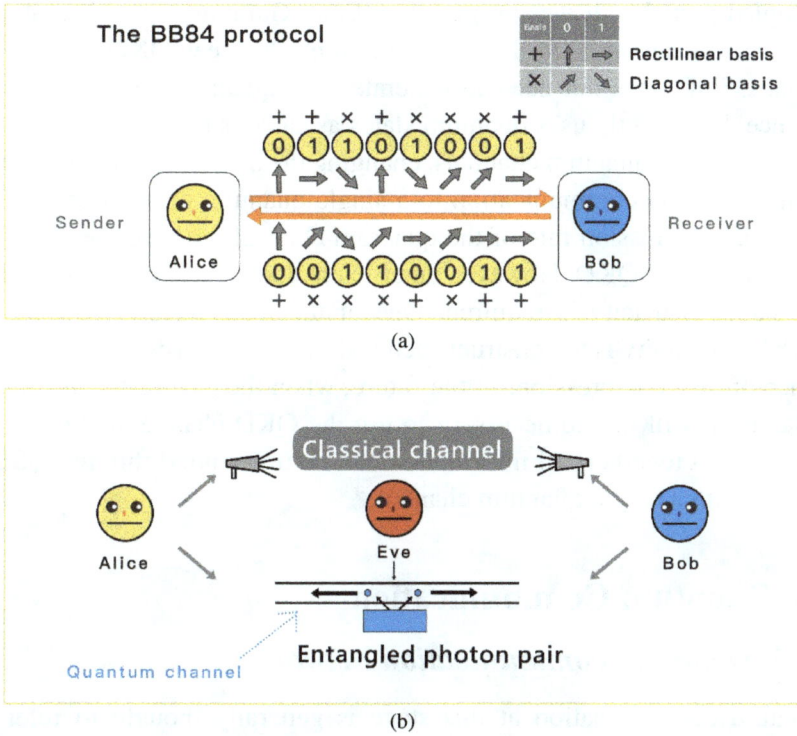

Figure 9.3. (a) Schematic diagram of the BB84 quantum encryption protocol. (b) Schematic diagram of the E91 quantum encryption protocol.

entangled state are correlated, any eavesdropping will destroy the entangled state between them. Alice and Bob can simply check Bell's inequality to find out whether Eve is eavesdropping, thereby ensuring the absolute confidentiality of quantum communication. The two parties can also compare the results through the classical channel. If someone eavesdrops, it will destroy the entanglement result and be easily discovered.

Another communication method that is completely different from QKD is Quantum secure direct communication (QSDC). The security of QSDC is also based on the principle of quantum non-cloning, the principle of quantum uncertainty, and the correlation and non-locality of entangled particles. It was proposed by Long Guilu and his doctoral student Liu Xiaoshu in 2000. The main difference between QKD and QSDC is that QSDC does not need to generate keys quantum mechanically in advance, but directly uses quantum states as carriers to encode and transmit messages in quantum channels, changing the dual-channel structure of traditional QKD communication to a single quantum channel of QSDC. The data transmission rate of the quantum channel required by QSDC is similar to that of QKD. However, an efficient coding scheme approaching the channel capacity is required. One of the major efforts now in the QSDC community is to construct such coding schemes with better performance of data compression. In the future, when the channel capacity can be larger, it will not be necessary to use the QKD channel and classical data channel together; all information can be transmitted through QSDC directly with a single quantum channel.

9.4 Quantum Communication

9.4 1 *Quantum communication network*

Quantum communication at this stage is generally thought to refer to QKD technology. However, the essence of QKD is not in communication but in generating a string of keys. With the emergence of various secure QKD protocols, there is also a need for a high-speed and long-distance quantum network. QKD and quantum networks are both integral components of quantum communication. Long-distance quantum entanglement transmission is the basis for building a global quantum communication network. However, since photons will be rapidly lost in optical fibre transmission, the construction of a quantum network needs to adopt the quantum relay scheme. In the quantum relay scheme, the task of long-distance entanglement transmission is decomposed into multiple short-distance linkages. Entanglement is established between quantum memories at multiple endpoints of these basic linkages, and then entanglement swapping techniques are used to extend quantum entanglement across a long-distance quantum communication network. The current quantum network can be divided into three directions: ground-based network, satellite–ground network, and space-based network.

(1) **Ground-based quantum network:** Due to the good compatibility of photons with optical fibre communication networks, photons are considered to be the best medium for quantum communication. However, since the loss of photons in optical fibre propagation increases exponentially with distance, the quantum communication distance cannot be too long. Generally, there will be a problem of a weak signal at about 100 km. Recently, Toshiba developed a new dual-band stabilization method, where two optical reference signals are sent at different wavelengths to minimize phase fluctuations within a fibre. The first signal cancels the changing fluctuations, while the second signal is used for fine-tuning the phase. This allows the optical phase of the quantum signal to remain within a small fraction of the wavelength, even in optical fibres exceeding 600 km. Traditionally, quantum repeaters have been considered the best method to overcome distance limitations by improving photon loss in optical fibres. Another possibility is to use satellites to send photons in a free space channel, with the current record being 1,200 km. Effort is needed to extend the propagation distance of photons in the communication network and enhance the efficiency of satellite-to-ground networking. In 2017, China built the world's first quantum secure communication trunk line, the Beijing–Shanghai trunk line. In 2022, the team led by Professor Long Guilu of Tsinghua University in Beijing designed a novel QSDC system that uses photonic time-bin and phase quantum states, realizing QSDC over 100 km fibres.

(2) **Quantum satellites with satellite-to-ground networking:** Long-distance optical fibre transmission is not only lossy but also leads to a decrease in the quality of quantum entanglement. Even if the best-quality optical fibre is used for transmission, after 1,200 km, only one photon can be transmitted every 30,000 years, and thus, long-distance quantum communication is impossible. On the other hand, quantum communication in free space using the ground surface is easily affected by weather and ground obstacles. Therefore, using satellites as photons for long-distance transmission in free space is the only viable option. When photons are transmitted from outer space through satellites, the loss can be reduced to one part in a trillion. With the current technology, one photon can be transmitted in one second, and

enough experimental data can be accumulated soon. In 2012, a team led by Professor Pan Jianwei from the University of Science and Technology of China successfully achieved the world's longest teleportation of quantum states in a 97-kilometre-long free space on both sides of Gangcha Lake in Qinghai, confirming the feasibility of quantum states passing through the atmosphere. In 2016, the world's first quantum science experimental satellite was launched, successfully demonstrating satellite-to-ground networking. Later, intercontinental quantum communication spanning a distance of 7,600 km from Hebei to Austria was achieved using the satellite.

(3) **Space-based quantum network:** In 2020, a Singapore team used nano satellites to demonstrate quantum entanglement in free space. The CubeSat nano satellite weighs less than 2.6 kg and is smaller than a shoebox. Aitor Villar said, "In the future, our system could be part of a global quantum network transmitting quantum signals to receivers on Earth or other spacecraft." In 2021, a team led by Professor Zhu Shining of Nanjing University in China published data on the transmission of entangled photons between drones, hoping to use them as nodes of mobile quantum networks to open up near-space-based quantum communication. China is also studying the possibility of using satellites as relays for ground-based quantum transmission. China launched mini-satellite Jinan 1 in late 2022, which conducts quantum communications in the lower Earth orbit. This is the first step towards establishing an ultra-secure communications network with global coverage. More quantum mini-satellite launches are planned to support future quantum communication in the sky. In addition to the ground-based network, the sky of the future quantum network will be covered with countless Quantum mini-satellites forming a three-dimensional global quantum communication network in which space-based networks and ground-based networks are connected through satellites and the ground.

9.4 2 *Repeaters used in quantum network*

Transmitting entangled qubits over long-distance fibre-optic networks is a huge challenge. The no-cloning theorem, while ensuring that eavesdropping is impossible, also makes it impossible for quantum information to be amplified like electrical signals. Optical fibre is susceptible to temperature changes and vibrations, meaning photons suffer substantial losses during long-distance transmission. Suppose that a photon source releases 10 billion photons per second. If the loss of photons per kilometre is one-twentieth in the optical fibre, after 500 km, the photon number per second will diminish to $(0.95)^{500}$ of 10 billion photons; not even one photon per second can reach the end. Sending 10 billion photons per second through fibre optics to a location a 1,000 km away would take around 300 years to receive just one photon. Therefore, long-distance quantum communication is not very practical. In classical communication, many signal-enhancing amplifiers can be installed along the way to overcome the loss problem, but classical amplifiers will destroy the quantum characteristics of photons. In order to achieve long-distance quantum transmission, an intermediate node similar to a gas station on a highway is needed — a quantum repeater. These repeaters are placed along the quantum channel, decomposing the long-distance high-loss transmission into many short-distance low-loss composite quantum channels connected by repeaters. The entire quantum network realizes the absolute security of end-to-end communication through quantum repeaters. The method uses the quantum repeater to realize the effect of transmitting the quantum entangled state. Because there is no distortion or measurement in a quantum repeater, it provides absolute security of communication. The current mainstream method replaces the quantum state of photons with other quantum states for temporary storage, and the full quantum network can also maintain the quantum state at the relay point, which is more secure. However, the disadvantage is that the quantum state of quantum memory can only be maintained for a short time, so quantum information cannot be transmitted over long distances. If the entangled state can be maintained for a long time in the future, the entire quantum network will be the safest information channel.

However, before a quantum repeater for an entire quantum network arrives, repeaters to relay quantum information are necessary to transmit entangled information. Yet, many technical bottlenecks in developing quantum repeaters have yet to find breakthroughs. Three main types of repeaters are currently under development and used: **quantum repeaters,**

trusted repeaters, and secure repeaters. Quantum repeaters and trusted repeaters support two different types of QKD networks: (trusted node network) and (full quantum network). Recently, a (quantum and post-quantum hybrid network) using both quantum direct communication and secure repeaters with classical PQC has also been proposed.

9.4.2.1 *Quantum repeater*

Quantum repeaters mainly realize the entanglement manipulation of quantum states through technologies such as entanglement swapping and quantum memory so that quantum information can be transmitted to a longer distance. Assume that the receiver and the sender each have a pair of entangled photons, and they each send out one of the photons, and the two photons are entangled in the quantum repeater halfway. This allows the photons remaining at the sender and receiver ends to become entangled with each other. However, it is impossible for the photon of the sending and receiving parties to reach the quantum repeater at the same time, so in the quantum repeater, there must be a quantum memory to store the quantum information that arrived first. The quantum memory does not make any measurement on the photon but simply stores it and waits for another photon to form entanglement swapping. Quantum memory enables repeaters to connect two adjacent quantum fibre channels, which can be gradually expanded to form longer-distance quantum channels until the entanglement can be shared between the two remote ends, Alice and Bob. The quantum memory is a key component of the quantum repeater, which is used to store the photon entanglement state. Once neighbouring quantum memories have successfully entangled, the next step of entanglement swapping can be performed. A reliable quantum repeater needs to have at least three functions, as shown below:

(1) Needs to use standard telecom wavelength photons.
(2) Needs to possess quantum memories to store and relay entanglement data.
(3) Needs to swap data between nodes in a network with a controllable way.

With an improvement in the performance of quantum repeaters, long-distance quantum communication may be truly completed in the near future, and then quantum Internet will also appear. Many laboratories are working to extend storage times and improve the efficiency of collecting photons, hoping to achieve the ultimate goal of long-distance quantum communication networks.

In 2021, the University of Science and Technology of China team compared the two separated quantum endpoints to "Cowherd" (牛郎) and "Weaver Girl" (織女). In the experiment, the Cowherd and Weaver Girl directly established long-distance entanglement with the help of a quantum repeater called "Magpie Bridge" (鵲橋). Before meeting each other, Cowherd and Weaver Girl independently created pairs of entangled photons, with one photon from each entangled pair transmitted to the "Magpie Bridge" for Bell's entanglement test. Each successful Bell state test represents the completion of a successful case of entanglement swapping. Quantum entanglement was successfully established between "Cowherd" and "Weaver Girl" separated by 3.5 m. In 2022, Delft University of Technology in the Netherlands published results in the journal *Nature*, demonstrating the teleportation of quantum information between two non-adjacent nodes in a three-node quantum network, which is an important step towards the realization of the quantum Internet. In 2023, research on diamond defects as quantum repeaters demonstrated that they can help quantum signals travel through 50 km, connecting Lincoln Laboratory, MIT campus, and Harvard University. Signal photons encoded in quantum states sent from Harvard University are connected to quantum networks, where the transmitted quantum states are captured and stored in quantum memory before being sent to the recipient. Furthermore, the findings suggest that quantum signals can travel up to the long distance required for practical applications. Another group from Innsbruck also shows that trapped ion-based quantum repeaters allow the transmission of entangled photons over 50 km.

9.4.2.2 *Trusted repeater*

The trusted node network connects many short-range QKD communication networks using trusted repeaters, and the short-range network between two trusted repeaters can directly use QKD technology. Keys can be exchanged between adjacent trusted repeaters, so each trusted repeater possesses the key. Therefore, every node within the network must be secure and trustworthy. Each trusted repeater can independently perform key generation and deliver the key to the client. Alice and Bob, who are far away at the final two ends of the network, can use quantum communication through multiple trusted repeaters. Before the quantum repeater technology matures, the trusted node network, combining the advantages of classical networks and quantum networks, can be used commercially, and it is currently in use in China and Japan. At each node of the trusted node network, the arriving photons are collected and measured, and then the photons of the correct state are resent to the next node in the network. As long as the network has enough trusted nodes, it can overcome the distance limitation of QKD transmission. Another benefit is that different types of QKD protocols can be used in the same communication channel. The main disadvantage of trusted node networks is that the transmitted photon needs to be measured at each trusted repeater, leading to the complete destruction of the original quantum state. As a result, it cannot provide absolute security for transmission. From the principle of quantum encryption communication, complete security can only be ensured within a communication distance of several hundred km because all relay points have the risk of information leakage.

Recently, China established a trusted node network between the Micius satellite and three ground nodes, enabling QKD over the satellite–ground network. Of the trusted repeaters on the ground, two are in China and one is in Austria. In 2017, scientists from Austria and China held an secure video conference using satellite-to-ground networking. Quantum communication in space has two major drawbacks. The first is low transmission rates, as only about one photon out of every six million photons sent from a satellite per second is measured on the ground. The second drawback is that the current quantum communication of Micius can only operate at night. Hence, the National Space Science Center of China hopes to launch satellites with stronger signals to operate during the day.

9.4.2.3 *Secure repeater*

The quantum and post-quantum hybrid network can construct a secure quantum network by using quantum direct communication and secure repeaters with classical PQC, that is, using quantum direct communication to transmit ciphertext, and building a large-scale network through classical relay. At the network node, the ciphertext has the protection of a secure repeater with PQC; thus, a quantum network with end-to-end security can be built. In short, PQC replaces the human at trusted repeaters in the network. The secure repeater network is a novel idea of constructing an end-to-end security hybrid network using existing technology, exploiting QSDC and classical cryptography. In a nutshell, it uses QSDC to transmit ciphertext quantum mechanically within fibres. At the network node, the ciphertext is read out, routed, and transmitted again using QSDC to the next node. In the node, the text is put through the repeater and relay in a classical way. However, it is protected by classical cryptography and, in particular, can resist quantum cracking if PQC is used. The idea was proposed by researchers from the Beijing Academy of Quantum Information Sciences and the University of Southampton, UK. The government of Beijing city has planned a priority project to build a demonstration network in its 14th five-year plan.

There are pros and cons for the three different repeaters mentioned earlier. They all provide the security of quantum communication and are currently being used worldwide.

9.4.3 *Quantum teleportation and entanglement swapping*

The principle of quantum teleportation divides the original object into two parts, classical information and quantum information, and transmits them to the receiver through the classical and quantum channels, respectively. Classical information is the information obtained by the sender after measuring the original object. Quantum information is the information the sender has not extracted during the measurement. Upon receiving these two kinds of information, the receiver can recreate a perfect clone of the quantum state. In this process, only the information about the object is transmitted, not the object itself. The sender may not even have any knowledge about the quantum state of the original object. According to the received information, the receiver adjusts the particles in his hand to the quantum state of the original object.

When the quantum state being teleported is an entangled state, quantum teleportation becomes entanglement swapping. Two originally unrelated particles can be entangled by entanglement swapping. Quantum teleportation and entanglement swapping can accurately transmit the quantum information of the original object to a distant place. This "spooky action at a distance" is similar to teleportation in science fiction films, but it cannot be used to transmit objects. It is possible to achieve long-distance QKD through quantum teleportation, thereby adding an absolutely secure "quantum lock" to quantum communication.

As shown in Figure 9.4, particles 2 and 3 are entangled with each other. After Alice gets the original object (particle 1), she performs a Bell state measurement on particles 1 and 2, and then transmits the measurement information (classical information) to the remote Bob through a classical channel. Since Alice and Bob have a pair of entangled quantum particles, 2 and 3, when Alice measures particles 1 and 2, entanglement causes the state of particle 3 to be determined simultaneously. Bob, using the Bell state measurement results obtained from Alice, can then recreate the quantum state of particle 1 on particle 3. This process is neither

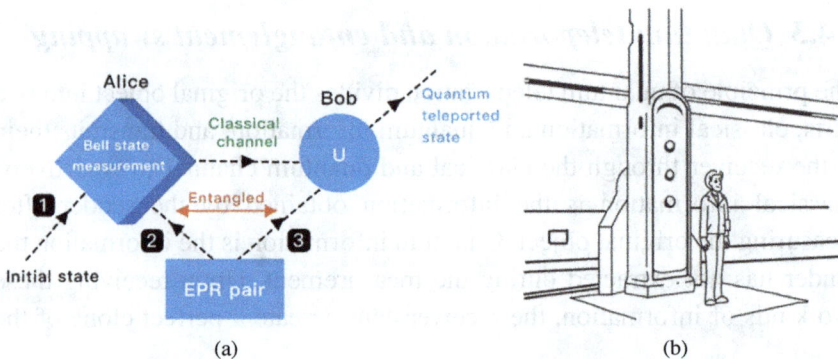

(a)   (b)

Figure 9.4.   (a) Quantum teleportation and (b) teleportation in science fiction movies.

cloning the original object nor cloning the quantum state of particle 1, but rather quantum teleportation of the quantum state. Alice's measurement result still needs to pass through the classical channel, so the transmission speed cannot exceed the speed of light. Therefore, quantum teleportation is not the instantaneous teleportation seen in science fiction movies.

9.5 Quantum Internet

9.5 1 *Anti-eavesdropping quantum Internet*

Quantum Internet refers to connecting various quantum components to transmit quantum information through quantum channels. Quantum Internet is an Internet that uses quantum principles such as quantum entanglement to transmit information. Leveraging the properties of quantum mechanics, quantum Internet ensures absolute security in network communication. As we all know, the Internet is a global system for transmitting, processing, and storing information, while quantum Internet is designed to transmit, process, and store quantum information. Qubits and quantum entanglement are fundamental elements of the quantum Internet. Scientists build quantum computers to solve specific problems more efficiently, not all problems. Similarly, the purpose of quantum Internet is not to replace the existing Internet.

In practical applications, QKD should be used in conjunction with cryptographic algorithms, and it should be through quantum channels and classical channels at the same time. Quantum channels transmit quantum signals to distribute quantum keys and codes, and classical channels transmit non-quantum information. However, for the quantum Internet to move from the laboratory to widespread use, two major challenges need to be addressed: the security problem under realistic conditions and the long-distance transmission problem. Through the academic community's efforts for many years, the practical safety distance of point-to-point QKD using optical fibres is about 100 km. Recently, Toshiba Corporation reported that it could reach 600 km, but practical implementation still requires further work.

With the existing technology, the use of trusted repeaters can effectively extend the distance of quantum communication. Recently, quantum repeaters also have made very promising progress. In 2017, the quantum secure communication trunk line, from Beijing–Shanghai, utilized 32 trusted repeater nodes to establish an intercity optical fibre quantum network spanning approximately 2,000 km. More importantly, it successfully connected to the quantum satellite Micius, becoming the first quantum Internet in the world. The entire network covers four provinces and three cities. In addition to the Beijing–Shanghai main line, it also includes four quantum metropolitan area networks in Beijing, Jinan, Hefei, and Shanghai. It is connected to Micius through two satellite ground stations, with a total distance of 4,600 km. There are already more than 150 users in industries such as finance, electric power, and government affairs. The Micius quantum satellite was recently used as a relay to further expand the 7,600-kilometre intercontinental quantum communication between Hebei and Austria in the free space channel. In 2022, another low-orbital quantum mini-satellite, Jinan 1, was also launched. More quantum communications ground systems in China are under construction now.

In August 2020, the US Department of Energy released the report "Building a Nationwide Quantum Internet," proposing a strategic blueprint for building a national quantum Internet within 10 years. The report defined a potential roadmap towards building the first nationwide quantum Internet and hoped to ensure that the United States was at the forefront of the global quantum competition, leading a new era of communication.

9.5.2 *Quantum Internet of Things (QIoT)*

The Internet combines computing, sensing, and communication functions, while the Internet of Things (IoT) is the automatic connection of various components through the Internet. The Internet connects information, whereas the IoT digitalizes and integrates the real world, optimizing digital information from all objects online. The applications of the IoT are extensive, encompassing various interactive and transactional patterns developed by humans over thousands of years. These can be digitally represented on the IoT, providing new opportunities and applications.

The IoT is an emerging field attracting attention, but security and privacy are the main factors that are questioned for the application of the IoT. Since potential hackers may appear in all nodes on every network, security has always been the fatal problem of the classical IoT. QIoT uses quantum technology to do the same thing as the classical IoT, combining quantum computers/simulators, quantum communication, and quantum sensors with quantum measurements to become a QIoT. The advantages of the QIoT are not only the use of ultra-sensitive quantum sensors and fast quantum computers but, more importantly, the fact that the QIoT has the function of anti-eavesdropping, which guarantees the security and privacy of the IoT.

9.5.3 *Build a quantum Internet*

The purpose of realizing the quantum network is to surpass the short-distance QKD, give full play to the potential of quantum communication, and eventually move towards the global quantum Internet. Quantum Internet combines different network topologies, including Trunk, Metro, Tree, and Mesh, as shown in the Figure 9.5. The quantum Internet can use quantum repeaters or trusted repeaters as relay links for long-distance optical fibre networks. The memory in quantum repeaters can be realized using different systems, such as neutral cold atoms, trapped ions, and diamond NV colour centres. All kinds of quantum information and even quantum entangled states can be directly distributed to online users of different nodes through the quantum Internet. In Figure 9.5, Alice is using barium ions to store the quantum state, while Bob is using the spin memory of the diamond NV colour centre. Alice can transmit her quantum state to Bob through quantum teleportation on the quantum Internet. In addition to the safe transmission of quantum information, the quantum Internet can also use quantum sensors and quantum computers to engage in quantum precision measurement, quantum digital visa, distributed quantum computing, etc. The quantum Internet has three main features: The devices connected to the network are quantum devices, the network transmits quantum information, and the network transmission method is based on quantum mechanics. At present, commercial quantum computers

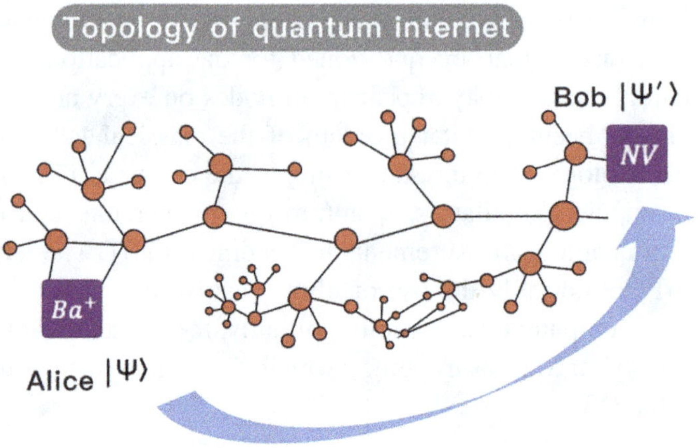

Figure 9.5. Trusted node network and full quantum network, different groups of dots represent the Trunk, Metro, Tree, and Mesh in different levels.

have yet to be formally applied, so the quantum Internet connecting quantum computers is still a future concept. The QKD quantum secure communication network currently being promoted by various countries is the prototype of the quantum Internet, and the ultimate goal is to integrate functions such as quantum computing, quantum sensing, and quantum measurement to form a quantum Internet. The United States plans to establish the basic capabilities of quantum interconnection, quantum repeaters, quantum memory, and even a high-throughput three-dimensional global QIoT. The future vision is that the quantum Internet can realize new functions that classical technologies cannot realize, and promote novel entanglement applications in human endeavours.

The US Department of Energy is experimenting with exchanging quantum information using a 30-mile-long underground optical fibre in a Chicago suburb. A number of research institutions in Europe plan to complete a network demonstration including 3–4 quantum communication nodes in the Netherlands in recent years, drawing a blueprint for a future European quantum Internet. The invention of the Internet has brought humanity into the information age, and quantum Internet will provide another opportunity to change the world. Quantum computers are expensive to build now. Early offerings will be limited to quantum cloud computing. Through the Internet, various services can be provided, allowing users to access quantum computers via the network to upload tasks and download results. Through quantum entanglement, many problems that classical computers cannot handle can be solved, such as entangling distant atomic clocks to improve the accuracy of time.

9.6 Concluding remarks

9.6.1 *Will the QIoT replace the IoT?*

Although the quantum information industry is still in its infancy, the concept of QIoT has already taken shape. China, the United States, the two major technologically advanced countries, and the European Union regard the quantum network and quantum Internet as long-term goals. The QIoT is the unification of all technologies in the "Second Quantum Revolution." It is a platform for quantum information transmission and storage, and an absolutely safe exchange channel for QC. The QIoT can also build a global atomic clock to redefine time synchronization, serving to achieve "Quantum Cloud Computing." Users anywhere in the world can be entangled with each other through the quantum Internet, and its various applications offer infinite possibilities and broad futures. The QIoT and classical IoT will be integrated into a hybrid system in the long run for their different applications and security levels.

9.6.2 *Future challenges of QIoT*

Quantum communication, quantum sensors, and quantum computing in the "Second Quantum Revolution" use many phenomena that are not found in classical physics, so they could also face many engineering and scientific difficulties. The QIoT is a comprehensive system that integrates all quantum components. The difficulties faced by the QIoT far exceed those encountered by each individual quantum component. The ideal QIoT is similar to the structure of the human body; it needs to have at least (1) a quantum brain, (2) a quantum sensory system, (3) a quantum transmission system, (4) quantum neurons, (5) quantum memory, and (6) unique entangled endpoints. The quantum brain is a universal quantum computer, which is currently only at the infant level and still needs to grow. Quantum sensors are used to prepare and measure various environmental conditions and parameters of network endpoints. Although considerable progress has been made, automation and sensitivity still need to be further improved. Quantum communication systems have proven effective over short distances and have been combined with trusted repeaters in China to extend the Beijing–Shanghai trunk line for more than 2,000 km. Quantum repeaters, similar to neurons, can spontaneously receive and transmit quantum signals, but there are still many difficulties. Quantum memory is an essential component, and the ability to store different quantum information at each node is crucial. Only the proof-of-concept outcome has been seen. The entanglement endpoint is the most powerful and novel function in the QIoT, and it has been proven to be feasible. But, if multi-point long-distance entanglement is to be carried out, all technical problems must be solved first. Only when these problems are solved will the real QIoT appear, which still requires a lot of effort.

The ideal QIoT in the future will entangle all quantum objects using quantum network, including connecting small and medium-sized quantum computers to form large ones. The giant quantum system integrated through the QIoT can perform various things that classical machines cannot achieve, for example, real-time simulation of quantum chemistry of molecules or materials. All new materials and new drugs can be designed in advance in quantum computers and tested, before moving to laboratory synthesis and factory production. There will be many applications beyond science in the future. For example, in elections, QIoT can allow voters to choose not only one candidate but also the superposition of candidates, including their second choice. Quantum voters can use strategic voting schemes that classical voters cannot implement. QIoT can also help large groups coordinate things and reach consensus, such as instant verification of complex systems like blockchain. Once all node users are entangled, administrative governance may undergo revolutionary changes. Can social systems be decentralized through trustworthy QIoT to achieve truly civilian governance of edge endpoints? All this may not be known until the advent of the unassailable QIoT.

Chapter 10

Quantum Education and the Future of the World

We must use quantum computing to benefit the future of the next generation.

— Ching-Ray Chang (1957–)

The future always comes too fast and in the wrong order.

— Alvin Toffler (1928–2016)

10.1 Quantum common sense and knowledge

Facebook founder Mark Zuckerberg once posted a photo on his Facebook page of him and his wife reading the toddler book "Quantum Mechanics for Babies" to their baby. This shows that fairy tales will also change after the advent of the quantum age. In an *Aesop's Fables* fairy tale, the smart crow drinks water that was initially too far to reach at the bottom of a pitcher by throwing small stones into the water. Through this story of a crow, this fable gave hope to people that when encountering difficulties, one should not give up but use one's ingenuity and creative spirit to make full use of one's surrounding things. However, the times have progressed and the tools that can be used have increased. Modern parents should tell their children that crows have to use straws (Figure 10.1), because using stones is like using a classical digital computer and using a straw is like

Figure 10.1. Modern crows no longer have to move stones one by one like using classical computers, but should use straws to quickly solve problems like using quantum computers.

using a quantum computer. The times have changed, and fairy tales have to keep up with the modern times.

10.2 Quantum early childhood education

The United States promotes quantum common sense in early childhood education from age three. Treating quantum phenomena as knowledge training is difficult because it is non-intuitive and complex. It is more effective for parents to use common sense as a bedside story when children are young, so *Quantum Mechanics for Babies* appeared. As children grow up, quantum games are used to continue attracting children to engage in various quantum concepts. This way, quantum common sense can be easily introduced with life-growth experiences. In 2019, the fairy tale *Qubits and Quiver Trees* was published in the United States for children over 5 years of age. Qubits are a new budding technology, and the Quiver tree, a unique aloe tree in South Africa, is an endangered plant. The book introduces 5-year-old children to the most interesting and important careers in the next 20 years. The storybook divides work into two kinds: qubit-like work and quiver-tree-like work. Qubit-like work makes use of future technology, while quiver-tree-like work focuses on the current occupations. Both are equally important, but at the same time, the book tells the child that occupations that do not currently exist will become important one day. Therefore, they should pay more attention to learning about future knowledge and technology and be careful not to fall into the occupation trap that is on the verge of disappearing.

The concept of quantum is very unintuitive. Quantum phenomena will not appear in daily life. However, once the era of quantum technology comes, everyone will live in the ecosystem with quantum products. Therefore, quantum literacy education will become extremely important in the future. Only citizens with quantum literacy can be successful in the quantum era. The dilemmas of quantum education include the following: How can one effectively and quickly equip laymen with quantum literacy? How can one allow managers interested in understanding quantum technology learn how to make correct judgments? How can we ensure that those who hope to become quantum computer experts have channels to participate formally?

10.3 Quantum games and quantum science fiction

10.3.1 *Quantum games*

Games are often the first real-world applications of any new technology, including blockchain and artificial intelligence. Games provide the best training for kids in hands-on operations. Gamers are wondering what quantum computers can do for gaming. This is an important question for the gaming industry, but at the same time, it can also be asked in reverse: What can games do for quantum computers? Rapid optimization is an advantage of quantum computing, and thus promises better quantum artificial intelligence. Artificial intelligence is an important factor for behaviour of non-player-controllable character in games, and it will be more challenging and interesting due to the participation of quantum artificial intelligence. It also means the ability to create more realistic, accurate, and detailed characters beyond what players encounter today. Similar to quantum computing, quantum gaming is still just getting started, but researchers and developers are already working to bridge the gap between theory and reality. The quantum games that have just appeared at present basically directly add elements of entanglement and superposition to some traditional games that are already well known. However, this method is not the best way to promote quantum games because creativity and fun are limited by simply adding quantum attributes. Moreover, players are already too familiar with the original game and lack a challenge and freshness. As of now, although many quantum games have appeared, there are still no new quantum games on the market that really arouse the interest of players. The following are some current examples of quantum games:

(1) *Qubit the Barbarian*: The maze puzzle game illustrates the basic concept of quantum computing: Unlike traditional computers with binary (1 or 0), quantum computers based on qubits can make different states simultaneously appear. The superposition state is the core of the tasks performed by quantum computers, and the execution speed of game tasks is much faster than that of classical computers. It is worth emphasizing that quantum computing has opened a completely different door for random number generation, which means that more unpredictable game maps and character encounters will appear in games.

(2) *Hello Quantum*: The mobile app developed by IBM Q and the University of Bath in the UK briefly introduces some operations of quantum gates and can be cleared in about 15 minutes. The rules of how to play *Hello Quantum* are easy to master, but it is not easy to become an expert. Through the game, one can learn some key behaviours in the mysterious world of quantum mechanics and further develop more quantum knowledge in the future. This mobile game is an app game designed for children over the age of four.

(3) *Quantum minesweepers*: The famous classical game of minesweepers is expanded by adding quantum concepts. The main purpose is to teach the peculiar concept of quantum mechanics in an interesting way. *Quantum minesweepers* demonstrates the effects of superposition, entanglement, and their non-local characteristics. In the classic *Minesweeper*, the goal of the game is to find all the mines placed on the game board without triggering the mine. But, in the quantum version, several game boards are in a state of quantum superposition, and

the goal of the game is to determine the precise placement of all mines in all superimposed boards. There are three types of measurements in the game: classical measurements, which collapse quantum superposition states; measurements without quantum interactions, which can probe mines without triggering them; and entanglement measurements, which provide non-local information.

(4) **Quantum Battleships**: Like the traditional game *Battleships*, the quantum version is played on a grid, with each dot representing a place where a ship might hide. In order for the ship to operate on a real quantum device, the only equipment used is IBM's five-qubit prototype quantum processor and quantum programmes written in IBM's open-source Qiskit software package.

(5) **Quantum tic-tac-toe**: *Quantum tic-tac-toe* is a quantum derivative of the traditional *tic-tac-toe*, which allows entry into the quantum world without complicated mathematics. It introduces quantum concepts through the rules of the game, aiming to cultivate quantum intuition. Three quantum phenomena are introduced in the *quantum tic-tac-toe* game: quantum superposition, quantum entanglement, and quantum collapse. Quantum collapse refers to the phenomenon in which a quantum state is reduced to a classical state, which occurs whenever a measurement is made. In traditional *tic-tac-toe*, one directly marks X or O in one square at a time, but quantum tic-tac-toe players must mark two squares of quantum entangled pairs at a time, not one. The pair of X or O is labelled with numerical subscripts counting from 1; the pairs of labelled X or O are entangled and called spooky pairs. When the game progresses to the point where all the marked entangled X or O form a ring-shaped entanglement connection in the tic-tac-toe square, an automatic quantum measurement is triggered, and all spooky pairs collapse directly to the classical result. At this time, it is like Schrödinger's cat in the opened box — only one of the entanglement marks can exist. All Xs or Os on a ring connection are then measured and uniquely determined, and can no longer be changed during the game. However, the measurement outcomes are not decided by the player who completes the entangled loop. Instead, his opponent decides on the final collapsing patterns. The winner of the game is determined in the same way as the traditional *tic-tac-toe*.

If, after the collapse, both X and O form a line simultaneously, the one with the smaller sum of the subscript numbers of X or O wins.

10 3.2 *Quantum science fiction*

Classic movies have a long-lasting and far-reaching impact on human culture, and sci-fi movies that involve infinite imagination often provide unexpected inspiration for the future of technology. In the 1980s, Michael Knight, the protagonist of "Knight Rider," drove the omnipotent Knight Industries Two Thousand ("KITT") to fight crime, which was popular among teenagers. Now, autonomous vehicles have become an actual product on the streets. In the past, many sci-fi films about landing on the moon, Mars, and interstellar travel also provoked curiosity about the unknown universe and inspired the vision of countless entrepreneurs who were willing to invest in the space industry to fulfil their childhood dream of traveling freely in the universe. Recently, SpaceX's founder Elon Reeve Musk, Amazon's founder Jeff Bezos, and Virgin Atlantic's founder Sir Richard Branson all invested in the development of space vehicles. The future of human space travel has gradually become an itinerary just around the corner.

Quantum sci-fi movies are a type of sci-fi movie with fantasy scenarios based on quantum science. The main appeal of science fiction films lies not in science itself, but to satisfy human fantasy, fill the emptiness of the soul, and even fulfil the dream of immortality! Quantum sci-fi movies often employ scientific theories that do not fully adhere to rigorous scientific facts, such as time travel and superpowers. Science fiction movies often use the future world as a story background to avoid challenges from scientists. Because many stories in quantum science fiction films have surpassed the capabilities of modern technology, we can only imagine what will happen in the future world. The reason why sci-fi movies easily attract audiences is that they can inspire unlimited imagination, and the plots of sci-fi movies are sure to appear in the future as alternative technologies. For example, sci-fi capabilities such as clairvoyance can be achieved by today's mobile phones. Therefore, science fiction films will also play an important role in quantum education to stimulate creativity and imagination. Quantum science has several features that are absent in the classical world and, thus, often appear in quantum science fiction films. There are many typical quantum science fiction films. The following only lists the most representative ones and the basic principles of quantum phenomenon associated with them:

(1) ***Star Trek* and quantum teleportation:** *Star Trek* first aired on the US TV channel NBC on September 8, 1966 and ran for three seasons thereafter. The story is about the interstellar adventures of Captain Kirk and the crew of the United Space Ship (USS) Enterprise (NCC-1701) in the 23rd century, which spawned animated series and several movies. One of the most striking ideas in the series is that of the teleporter, a common form of short-distance travel in *Star Trek* that breaks down a human body or matter into quanta and sends the quanta to the destination to reassemble. The teleporter in *Star Trek* was originally an idea that the production staff came up with to save the cost of filming, as the cost of filming landing scenes on an alien planet was much higher than that of the teleporter. In *Star Trek*, when Captain Kirk wanted to return to the Enterprise from a distant planet, he always asked Chief Engineer Scott to "Beam me up, Scotty." The teleporter phenomenon is only seen in sci-fi movies or magic shows. However, the idea of a teleporter is somewhat similar to quantum teleportation at a certain level, except that quantum teleportation can only transmit information, not an object. However, "Beam me up" became the teleportation mantra. This teleportation trick was later used in many sci-fi films, even films involving teleportation across time, as in the *Back to the Future* series.

(2) ***Tenet* and the reciprocal process of entropy reduction:** *Tenet* is a sci-fi action thriller that was released in the UK and the US in 2020. It proposed several sci-fi elements, incorporating the concept of "temporal pincer movement." This film is an innovative, brain-burning masterpiece by the great director Christopher Nolan. Without a certain level of scientific knowledge and a very focused mind, it is not easy to understand the movie's complex storyline while watching it. Not only has it sparked heated discussions on the Internet about the various plots in it but, in addition to the grandfather paradox common in sci-fi films, *Tenet* also showcased plotline that constantly

reciprocated between multiple time channels, which is impossible in the real world. One unique sight in the movie is when Barbara gives the protagonist an empty pistol and asks him to aim at a boulder full of bullet holes in the distance in the shooting range. The pistol magically recovered the bullets. This reciprocal process was explained as a "reverse time substance" that reverses entropy when exposed to specific radiation. Entropy indeed can be possibly reduced locally but not globally, and Christopher Nolan applied this idea in *Tenet* to indicate that reciprocal processes are possible in his science fiction world. The plot is so complex that even the actors were sometimes unsure about what they were filming each day and only gained a preliminary understanding after the film was edited. "Don't try to understand it, just feel it" may be the mentality that the audience should have when watching it. In fact, when ordinary people come into contact with quantum technology, they need to hold a similar mentality: "Don't try to understand it, just accept and use it." While there are many science fiction films that involve time loops, Christopher Nolan's *Tenet* sets a new standard by incorporating time loops within multiple spaces. The movie name *Tenet* also expresses the idea of reincarnation and repetition.

(3) ***Ant-Man and the Wasp* and quantum tunnelling:** The character "ghost" in *Ant-Man and the Wasp* undergoes a transformation after an accident so that she can pass through various objects. The movie tries to rationalize why the "ghost" can tunnel through any object; this is because after her accident, she has become a state of quantum probability. Because it is a probability wave, the wall-penetrating phenomenon of quantum tunnelling can occur. However, according to the wave–particle duality of quantum physics, this cannot happen in the macroscopic world because the matter waves of human-sized object cannot be observed. In science fiction films, wall penetration is common, which is also an audience favourite. In fact, the ancient Chinese story of *Liao Zhai*《聊齋》also mentioned that Mr Wang and Taoist priests in Laoshan Taiqing Palace (嶗山太清宮) learned the wall penetration technique. Even today, when one visits Laoshan Taiqing Palace, there will be tourist guides who show the wall that was supposedly penetrated in the past. Magicians and sorcerers also like to

confuse ignorant people by supposedly showing them pills that pass through bottles and objects that come out of walls.

(4) **Telepathy and quantum entanglement:** *Lucy* was released in 2014. It is a French science fiction action film, mainly set in France and Taipei. After the lead character Lucy accidentally takes a magical drug, her brain function rapidly evolves and she is able to read other people's minds. This kind of telepathy is also commonly seen in other sci-fi films, such as the unique telepathic ability of the Vulcans in *Star Trek*, which enables them to communicate with others by touching their faces and sharing their consciousness, experience, memory, and knowledge.

Telepathy is the phenomenon or ability to transmit messages to another person at a distance. But, so far, no one has been able to prove its existence using rigorous scientific methods. Some people like to connect quantum entanglement with telepathy, mainly because quantum entanglement is a long-distance spooky effect. After a quantum measurement is made at one end, the results will collapse together, which is somewhat similar to the basic essence of telepathy. However, quantum entanglement is a controllable and reproducible scientific phenomenon that is derived from rigorous science, which is completely different from telepathy that cannot be reproduced as one wishes. However, science fiction movies still often like to portray this kind of strange ability, and the preference for such science fiction plots also reflects the outline of the future world that humans expect. With the development of science and technology, scientists are trying to develop an artificial intelligence brain chip that can be implanted in the human brain. The idea is that even if two people are far apart, telepathy can occur between them through wireless communication, which hopefully will happen soon if Elon Musk's artificial intelligence brain chip proves successful. Maybe one day, the telepathic superpowers in science fiction movies will actually appear in the world by means of this new type of brain chip. When the technology matures, the sci-fi world of *The Matrix* may also appear in the real world.

Transcendence is a 2014 sci-fi movie. In the movie, scientist Will and his wife work on developing artificial intelligence and quantum

computers to build immortality technology, and finally have a breakthrough to create a quantum computer that surpasses the human brain.
After Will dies, his wife uploads his consciousness into a quantum
computer, and Will miraculously reconnects to the computer network
and is reborn. When Will possesses great wisdom and ability and
becomes an omnipresent, omnipotent, and unstoppable man-made
god, nothing can stop him. The plot ends dramatically where love
transcends all. The creation of *Transcendence* was inspired by a futur-
ist Ray Kurzweil. Kurzweil believed that immortality was the inevitable path for mankind: The first step involves using drugs and
supplements to maintain longevity. The second step takes advantage
of the genetic revolution to remain disease-free. The third step is the
nano-revolution — implanting nano-machines into the human body to
replace some organs and prolong life. Kurzweil predicted that by
2045, with the help of quantum computers and artificial intelligence,
humans will evolve into a hybrid of biology and machinery, thereby
gaining wisdom and immortality. At present, it seems that technologies such as quantum computers, artificial intelligence, and brain chip
integration are developing rapidly, giving the dream of immortality a
chance to gradually come true in the future.

(5) ***Blade Runner 2049*, quantum memory and consciousness:** *Blade
Runner 2049* is the sequel to the 1982 movie *Blade Runner*. In both
movies, artificial humans were created through bioengineering, and
artificial humans can be implanted with virtual memories. However,
in the movie, real humans generally believe that "artificial consciousness is not a real soul." *Blade Runner* attempts to think from an artificial person's perspective but wishes to live beyond humans. The
artificial man in *Blade Runner* felt the helplessness of ordinary life in
the pursuit of greatness, while the artificial man in *Blade Runner 2049*
pursued the living conditions of ordinary humans and was further
inspired by the nature of love input by artificial consciousness.

These science fiction movies discussed the origins of memory and
consciousness and where they are from. Memory and consciousness
are some of the most discussed issues since ancient times but without
any consensus till now. The relationship between consciousness
and matter, and how consciousness is formed, was never clarified.

Roger Penrose and Stuart Hameroff proposed a model based on quantum mechanics, which claimed that the brain's neuron system forms a complex network in which particles are entangled with each other. This theory has been greatly challenged by scientists and cannot be supported by any experiments. However, it has opened up discussions on the relationship between quantum entanglement and consciousness and sparked interest in countless science fiction films and pseudoscience.

(6) ***Everything everywhere all at once* and multiverse:** With its special effects and seemingly nonsensical narrative techniques, this movie won seven awards at the 95th Academy Awards, including Best Picture, Director, Lead Actress, Supporting Actor and Actress, Original Screenplay, and Editing, and became the first film to sweep the 2022 Academy Awards. The director applied the concept of multiverse from the high-dimensional Hilbert space spanned by many qubits. The movie tells the life stories of many Asian immigrants whose parents have high expectations for their children. They hide rebellion in their hearts. The heroine, Evelyn, who had a conflict with her father and gave up everything to elope, always disliked her rebellious daughter and weak husband. Evelyn travelled through the multiverse and always felt that "if … maybe I would have been happier, more accomplished, more …" But, in fact, it is the decisions we have made that have led us to become who we are now. Instead of regretting why we did not choose another path, we should focus on the present, cherish what we have, find our self-worth and strengths, face the environment with gentleness and kindness, and go all out to live life without further regrets.

10.4 Quantum education

10.4.1 *University education*

University education is where future industrial talents will be cultivated. At present, the world is actively developing quantum technology and hopes to establish quantum industrial parks. One of the essential conditions for the formation of quantum valleys is to have enough quantum talents. The US intends to train 1 million quantum engineers within 10 years and Japan hopes to have 10 million quantum programmers by 2030. The most important place for the cultivation of quantum technology talents is the universities; therefore, the university education system around the world has begun to change. Since quantum technology is comprehensive knowledge that spans multiple disciplines, it requires basic sciences, technology, engineering, arts, and mathematics (STEAM). At the same time, its application requires the integration of various professional fields of knowledge. Due to the complexity and diversity of knowledge involved, new disciplines must emerge to train first-class quantum technology talents. The education system around the world is also changing because of quantum technology. Many world-renowned universities have established quantum technology departments and institutes since 2021. For example, Harvard University recently launched a doctorate programme in quantum technology. In China, Tsinghua University recruited college students to study in the Andrew Yao Quantum Information Class. The University of Science and Technology of China also receives quantum undergraduates. University of New South Wales in Australia established the Department of Quantum Engineering for undergraduates. All these examples show the urgency of quantum education in universities. To meet the demand for quantum talents in the new era, China's Ministry of Education has also allowed universities to set up undergraduate programmes for quantum information technology. The era of quantum technology has come and is enthusiastically calling on young people around the world. Generation Quantum is about to emerge on the new quantum stage.

10.4.2 *Introductory books on quantum technology*

The following are a few basic introductory books on quantum technology:

(1) *Quantum Computing for Babies* (**Baby University** series) **2018:** This is a brief introduction to the application of quantum computers. In this book, babies, as well as adults, will discover the difference between classical bits and qubits and how quantum computers will change our future.

(2) ***Computing with Quantum Cats: From Colossus to Qubits* (Prometheus) 2014:** In the book written by John Gribbin, he first reviews Turing's work on the Enigma and the first electronic calculator, and then explains how quantum theory enables quantum computers to operate both in principle and in practice. The book goes beyond an introduction to theoretical physics and extends to explore practical applications of physics, from machines that learn by intuition to laptops and smartphones. The book further states that in the future, it is expected that quantum communication networks will really operate.

(3) ***Quantum Computing for Everyone* (The MIT Press) 2019:** This book is an introduction to quantum computing, explaining topics such as qubits, entanglement, quantum computing, and quantum teleportation. Chris Bernhardt wrote a book that anyone who can do high school math can read. After reading it, one will understand that quantum computing is not just a completely different discipline from classical computing but also the basic form of computing.

(4) ***Quantum Computing as a High School Module* (arxiv) 2020:** This book introduces the basic concepts of quantum mechanics required to understand quantum computing. This module introduces three key principles that govern how quantum computer works: superposition, quantum measurements, and entanglement. The goal of the instructional modules is to inculcate an interest in technology in high school students through hands-on work. This is a one-week high school teaching programme for school students aged 15–18.

(5) ***Dancing with Qubits* (Packt Publishing) 2019:** This quantum computing textbook comprehensively describes classical computing and the mathematical foundations needed to understand concepts such as superposition, entanglement, and interference. It also clearly introduces quantum circuits and quantum algorithms, explains the physical and engineering principles behind building quantum computing hardware, and finally explores the industries where quantum computing can be applied. In addition to introductions and explanations, each topic is presented with examples for those who want to delve deeper into quantum computing. This book requires a solid foundation in mathematics and is best suited for those with a strong interest in mathematics, physics, engineering, and computer science.

(6) ***Quantum Computing since Democritus* (Cambridge University Press) 2013:** The book introduces some basic thinking in mathematics, computer science, and physics. It begins with the ancient philosopher Democritus and goes through explanations of logic and set theory, computability and complexity theory, quantum computing, cryptography, quantum states, and quantum mechanics. Filled with insights, arguments, and philosophical perspectives, the book also covers astonishing topics such as time travel, Newcomb's paradox, the anthropic principle, and discussions of Roger Penrose's very peculiar ideas. Written by Scott Aaronson, winner of the 2020 ACM Prize in Computing, the informal scholarly style of writing will appeal to readers interested in the history of science and philosophy. Researchers in physics, computer science, mathematics, and philosophy will also benefit from this book.

(7) ***Programming Quantum Computers: Essential Algorithms and Code* (O'Reilly) 2019:** This book is a study manual for quantum computing programming. It is not an introduction to mathematics and quantum theory, but a direct demonstration of the unique capabilities of quantum computing. The book is divided into the following three parts:

  (i) *Programming for a Quantum Processing Unit (QPU)*: Explores the core concepts of quantum programming, including how to operate qubits and perform quantum teleportation.

  (ii) *QPU Primitives*: Introduces algorithms and techniques, including amplitude amplification, quantum Fourier transform, and phase estimation.

  (iii) *QPU applications*: Explains how to use primitives to build practical applications, including quantum search and Shor's factorization algorithm.

(8) ***Quantum Machine Learning with Python* (Apress) 2021:** This book covers the fundamentals of quantum mechanics and quantum computing, along with the relevant mathematical concepts. It also introduces how to use quantum computing to handle various application cases. It also covers quantum deep learning, which can assist in forecasting, financial modelling, genomics, cybersecurity, distribution network logistics, and cryptography.

In addition, there are also more advanced quantum technology-related books, and interested readers can check online to find resources on their own.

10.4.3 *Quantum programming competitions*

In recent years, companies including IBM, AirBus, BMW, and Origin Quantum have often held global quantum programming competitions. This kind of competition provides a good international platform for beginners to learn and get to know their peers. The business community proposes real-world problems, and participants from around the world come up with possible solutions. Through the competition process, people brainstorm and encourage each other to get the best results, achieving a win-win situation for academia and industry. More importantly, these competitions help cultivate talent in practical quantum programming and advance quantum algorithms. In international quantum computing competitions, the practical application of quantum programming is designed on various quantum hardware, and the latest quantum computing and application topics are discussed with outstanding international quantum computing experts. Young people need a suitable place to gather and practice their quantum abilities. Quantum programming competitions can provide the best motivation and stimulate all potential users. The cumulative results of open-source quantum programming codes also widen the scope of quantum applications.

10.4.4 Q-12 *Quantum secondary education*

The US National Science Foundation (NSF) and the White House Office of Science and Technology Policy are actively working with industry and academia to establish a National Q-12 Education Partnership to provide training tools for middle and high school students. Projects are already underway and are expected to expand quantum education across the US. The US government is committed to working with US educators over the next decade to establish a complete quantum learning environment, from laboratory tools to quantum development resources, and to build a complete pathway for quantum career development. With materials of quantum technology co-designed by teachers and scholars at all levels, teachers in the classroom will introduce students to quantum technology and the future of quantum careers through lessons and hands-on activities, nurturing the next generation for the United States to compete in the future quantum industry.

NSF promotes quantum education, including the establishment of Q2Work program to support teachers and students in quantum information science and technology. The main direction of implementation is to train basic workforce in quantum computing through interdisciplinary teaching of quantum computing in high school, to use summer seed teacher workshops to deploy quantum volunteers in the community, to organize various quantum promotion activities, to provide quantum teaching materials for middle and high school teachers to enable them to understand quantum information science and technology and effectively teach students quantum knowledge. In addition, high school quantum computer textbooks specially prepared for those aged 15–18 have been implemented for many years, and the results of the trial teaching have been compiled into reports and published.

In the planning of quantum education, currently only the United States has a more forward-looking and comprehensive approach in the world. From quantum fairy tales for babies to quantum games for young children and then to quantum sci-fi films and video games for teenagers, it comprehensively cultivates quantum common sense that citizens of the post-Silicon quantum era should have. Beyond high school, the integration of Q-12 training and university-level quantum education forms a complete framework for laying out quantum knowledge, aiming to train scientists and engineers needed for the quantum industry. Compared with the United States, the promotion of quantum education in other countries still needs to be strengthened. After all, future quantum engineers need high-quality training, and a long-term and stable source of quantum talent is an essential factor for success.

10.4.5 *Quantum ethics*

Quantum technology is rapidly developing from the late 19th-century knowledge into a huge industry. Many people worry that quantum technology will be abused. They compare quantum technology to the Manhattan Project during World War II. This is a new industrial revolution and this power, in the hands of the ambitious, could be used to create harmful substances or manipulated in dangerous ways. Whichever country can master key technologies first will possibly control the entire world. Although quantum computing still needs some time to achieve quantum supremacy, almost every new application of quantum technology is a destructive intrusion into the original system. If we do not start to consider ethical issues now, it will be too late when quantum advantages emerge. Also, because of the amazing power of quantum technology, some people urge the quantum industry to discuss the enforcement of quantum ethics in the early stage of development. Just like artificial intelligence and biotechnology, the world's ethical norms should be set as soon as possible. At present, quantum computing is being carefully promoted with the tangled psychology of both expectation of success and fear of harm. The quantum technology community has also noticed the urgency of the problem and made an urgent call. The first step is to raise social awareness, establish an effective legal and ethical framework, and limit the research and development content of identified risks within the scope of controllable losses. But, how do we combine legal, moral, and social aspects in the real world while encouraging quantum technology innovation? General guidelines for human-centred quantum technologies must be set by integrating the ethical, legal, and social implications (ELSI) associated with nanotechnology, artificial intelligence, and biotechnology.

10.5 The new quantum world

10.5.1 *The world quantum race has already begun*

Hollywood produced a sci-fi blockbuster in 2018, *Ant-Man and the Wasp*, which narrated the influence of quantum technology. The film's terrific dialogue between the villain's boss and the heroine is perfect to describe the future implications of quantum technology:

> "You think that I don't know what you've been building? With all of this? Quantum technology. And you can forget nano-tech. Forget A.I. Forget cryptocurrency. Quantum energy is the future. It's the next gold rush. I want in."

Hollywood's sensitivity to future technology is usually very accurate. This classic line clearly declares that quantum technology is the foundation of future technology and the correct goal of the future post-Silicon Valley era. There was also another movie, *Ant-Man and the Wasp: Quantumania*, in early 2023 in which Hollywood tried to show the world what quantumania is in future.

When the classical computer first appeared, Thomas Watson, the chairman of IBM in 1943, believed that the whole world only needs five computers. He once said "I think there is a world market for maybe five computers." In the 1960s, IBM invested about $5 billion to develop the mainframe IBM System/360, nearly three times IBM's annual revenue at the time. At present, IBM's annual revenue is about 74.5 billion US dollars, which means that even if it invests 220 billion US dollars into developing a quantum computer, it will be worthwhile. The total global investment in quantum computing is currently far less than this figure.

As quantum technology's future impact becomes increasingly apparent, developed countries have declared their entry into the quantum technology competition. For example, the United States passed the National Quantum Initiative Act (NQI), the European Union launched the Quantum Flagship Kickoff Meeting, Japan promoted the national plan Q-Leap, and China also declared that quantum technology is one of the main focus of the 14th Five-Year Plan. Quantum technology has sprung up like mushrooms after a rain, blooming all over the world. The quantum ecological environment has already taken shape. From the key technologies developed by multinational companies to the novel ideas of start-ups, the two legs of quantum technology and quantum industry are making great strides at the same time.

With the active support of the venture capital industry in the past few years, quantum computer-related technologies and applications have risen rapidly in recent years. There are at least hundreds of different types of quantum technology companies worldwide. If 2018 was the first year of the "Second Quantum Revolution," then 2021 was the first year of the commercial quantum company. The success of DWaves, IONQ, and Rigetti's IPOs in the United States has brought strong confidence in commercial quantum computers and attracted many new investors who hope to become technology upstarts. The quantum technology market is already involved in a gold rush. National resources, venture capital, and manpower are rapidly investing in this hot emerging technology market. It will soon be clear where the next quantum valley will appear. Chicago recently declared that it will become the Quantum Valley of the United States, replacing Silicon Valley as the new-generation technology centre. Waterloo, Canada, is also promoting its own Quantum Valley. Quantum Delta in the Netherlands, Quantum Avenue and Quantum Park in Anhui, China, and Quantum Valley in Munich and Lower Saxony, Germany, all demonstrate the promising future of quantum technologies. All of these developments showcase the strong global desire to form quantum industry hubs and compete for world resources in this field.

The development of the quantum industry in the United States is in full swing, and quantum technology companies are actively recruiting and cultivating talents. Quantum computing has become a complex cross-disciplinary engineering problem that requires a huge number of software and hardware engineers. At present, China urgently needs the government and enterprises to invest more resources so that it will not fall behind. The government should come forward to train a large number of talents, determine the future development direction, and establish a national quantum R&D centre. The key technologies in quantum computers, the development and manufacture of low-temperature qubit chips and control systems, rely on the cooperation of many emerging industries. Furthermore, the development of the quantum computer supply chain is also an important trend in the future, offering China the opportunity to evolve into the world's main controller of the quantum production chain structure.

10.5.2 *East wind presses west wind or west wind presses east wind*

One of the main axes of active competition between China and the United States in recent years is quantum technology. The European Union countries have also begun to gather forces and actively participate in the war. It remains to be seen who will end up winning in the battle of quantum technology. However, since the influence of quantum technology is strong, once successful, the existing technology is like an egg hitting a rock, and it may not be able to withstand a quantum blow. On the one

Figure 10.2. The fight between quantum computers and quantum communications is like a competition between spears and shields. As depicted in *Dream of the Red Chamber* (紅樓夢) by Cao Xueqin of the Qing Dynasty, "In all family matters, either the east wind suppresses the west wind, or the west wind suppresses the east wind." How the Eastern Shield will evolve against the Western Spear remains to be seen.

hand, people look forward to appearing as soon as possible to help solve problems such as new materials and new drugs, but on the other hand, they are also worried that the development of quantum technology will be too successful, resulting in an imbalance of international power and the emergence of an invincible force that controls the world. The current situation is that China and the United States are already the leading technological powers. China leads the world in quantum communication and the United States is the leader in quantum computing. The quantum computing that the United States is good at is like a spear, the quantum communication that China has an advantage in is like a shield (Figure 10.2), and the quantum sensor is extremely sensitive; because of its wide range, each has its own strengths and weaknesses. The spear has superior attack power and the shield provides perfect protection, but sensitive quantum sensors are still needed to effectively use the spear and shield.

In the future, we will enter the era of quantum technology. Quickly promoting quantum education among schools and the general public at this stage of rapid technological change requires the early participation and resource investment of different social classes at all levels. It is crucial for the next generation to enter the quantum ecological environment as early as possible. This century is the era of generation Q, and quantum education must start taking root early. All decision-makers must have a forward-looking quantum strategy; otherwise, when the era of quantum supremacy comes, they will be caught off guard. The rapid growth of the quantum technology industry over the past few years has not only been exciting but also terrifying. The exciting thing is that technology will create a better vision of life. The frightening thing is that if the technology of quantum computers is mastered by ambitious people first, the various existing security mechanisms of the current classical computers will be vulnerable, and the world order will be in chaos. What is even more worrying and surprising is that the British Ministry of Defence has recently purchased quantum computers for the development of smart tanks and smart fighter jets, which opens up another terrifying application field for quantum machine learning. The possible applications of quantum technology in the military field will effectively improve the accuracy of surveillance, target acquisition, network, space, electronics, underwater warfare, etc., leading to "quantum warfare." Therefore, the establishment of new military strategies, doctrines, and policies becomes necessary and also is an ethical challenge. Because of this, leading decision-makers need to pay close attention to the development of quantum technology and make necessary investments immediately.

10.6 Concluding remarks

Although some people have reservations about the competition of quantum technology, and even suspect that it is a replica of the Strategic Defense Initiative (SDI) (nicknamed the "Star Wars program") between the United States and the Soviet Union in the 1980s, suggesting it may just be a means of American war strategy. However, the main difference is that the Star Wars program was only a technological competition between the United States and the Soviet Union, while the quantum technology competition has obviously attracted the participation of governments and companies around the world, and its impact is not only in military facilities but also in applications in civilian production. Quantum technology will rapidly advance human civilization in the next few decades, and quantum philosophy will once again impact new thinking in humanities. Although social scientists still disagree on how to combine quantum theory with social science, there is also consensus. In the past, social sciences were mainly based on classical worldviews, and in the future, they may need to be considered and reconstructed in accordance with the thinking of quantum theory. In quantum social science, people may never be completely separate, but rather entangled social elements. Whether through language or other mechanisms, people are also caught in the dilemma of collective entanglement, just like particles in the microscopic world. Furthermore, the order of quantum measurements can affect the observed results. The corresponding phenomenon in social science theory is the order effect, where the results of a questionnaire are closely related to the order of the questions. Moreover, human thinking is never black and white, but resembles a superposition state in quantum theory. It often oscillates between "the Good in Nature" and "the Evil in Nature." It can even be like Schrödinger's cat before opening the box — no one will know the outcome beforehand. This Schrödinger's cat phenomenon of good or evil can possibly echo the Buddhist ideas of "turning back to the shore" and "putting down the butcher's knife to become a Buddha."

As modern quantum citizens, we do not have the right to reject quantum-related common sense. On the contrary, we should actively cultivate quantum literacy and bravely face the new wave of quantum. The "Second Quantum Revolution," which is in full swing around the world, has been dubbed by some people as a modern world war without gunfire. The research and development of quantum computers is like the Manhattan Project during World War II, but with laboratories replacing the battlefield, guns being substituted with knowledge, and scientists and engineers replacing warriors. However, the outcome of the tech war may still have a major impact on the world! After 2030, it is widely expected that quantum advantages and quantum supremacy will gradually appear in all fields, and life in the world will be happier due to the maturity of quantum technology. The US and China now try to maintain their respective leads in the post-Silicon Valley race. On the other hand, this is also a rare opportunity for all countries in the world, because everyone has just started to run from the starting line. While devising a correct quantum development strategy is challenging, relinquishing the opportunity to compete would be the most erroneous choice.

Quantum science was determined at the 5th Solvay Conference. The first important application was the Manhattan Project, which combined relativity and quantum theory. At that time, the whole world was trying to make an atomic bomb. Germany was led by Heisenberg, while Japan was led by Arakatsu and Nishina; but, in the end, it was the United States that reached the goal first. After the success of the Manhattan Project led by Oppenheimer, the US dropped *Little Boy* and *Fat Man* in Hiroshima and Nagasaki, respectively, to end World War II. Soon after World War II, the Soviet Union and the United States entered the Cold War period, and the semiconductor revolution was also in full swing. American scientists first understood the quantum confinement effect and quantum tunnelling effect, and made the MOSFET and CMOS. Therefore, the US took full advantage of the key technology of classical computers and maintained the world's leading and controlling position in semiconductor technology. The world entered an information era many years ago. The Internet of Things, which is composed of high-speed computers, sensors, artificial intelligence, and the Internet, has been woven into a global network. All human beings live on this invisible network, and their lives are deeply affected by the network at all times. In 2018, the EU officially formed a quantum flagship R&D team. It is hoped that after the "Second Quantum Revolution," the European quantum fleet can take the lead in the quantum era, continue Europe's glory of developing quantum science in the early 20th century, and let the achievements of quantum technology be carried forward in Europe. Europe can weave another quantum Internet of Things with quantum superposition and quantum entanglement. Over the past 100 years, the knowledge centre has moved from Europe, which was the source of quantum science, to the United States, which is the centre of world's economy, culture, and technology. At present, the world's invisible technology war is still in progress. Will China or the US gain a dominant position in the emerging quantum valley era in the post-Silicon Valley era? In the future quantum race, both poor and rich countries will stand together at the same starting line. How will the centre of the world move in the future? It remains to be seen who will win the battle of the "Second Quantum Revolution," but the only certainty is that active participation will provide better opportunities.

The quantum future holds endless possibilities. This century will be the quantum age. After the quantum concept emerged in the West and now enters Eastern culture, where quantum supremacy will go? As described in "Dream of the Red Chamber," in the struggle between the east wind and the west wind, which side will be the champion? The quantum era started its rise after classical technology and will be the next big thing to change the world following the industrial revolution. As Google scientist Neven said, "It is not one company versus another, but rather, humankind versus nature or humankind with nature." From now on, all countries should work together to make the earth a better place for all mankind.

Appendix A

Current Status of Quantum Technologies Worldwide

A1. Canada

Canada is one of the world's leading countries in quantum research, funding quantum computing since 1994 and investing more than US$1 billion in quantum research. The University of Waterloo established the Institute for Quantum Computing (IQC) in 2002 and has continuously expanded its scale in recent years to become an important quantum-related technology research institution in Canada. In addition to basic and applied research on quantum technology, D-wave, a well-known quantum annealing computing company, is also located in Canada. Waterloo claims to be the quantum valley and is at the epicenter of the quantum revolution, with 16 companies and over 250 researchers specialising in quantum cryptography, software, communication, and consulting.

A2. European Union

The EU held a symposium on quantum technology and industry in 2015, issued a quantum declaration in 2016, and launched a quantum technology flagship plan in 2018. The EU's future emerging technology axis plan in "Horizon 2020" includes quantum computers, quantum computing, quantum communications, and quantum sensors, with the aim of maintaining Europe's world leadership in quantum research. The European Union has also established a Quantum Internet Alliance, led by Delft University, which plans to establish the world's first optical fibre quantum communication experimental network between four cities in the Netherlands, and plans to build a global Quantum Internet made in Europe. The European Commission urges EU member states to cooperate in developing the EU's first quantum computer, improve their ability to independently develop quantum technology, and reduce external dependence. To unlock the transformative power of quantum, the EU should develop a solid industrial base that builds on its tradition of excellence in quantum research.

A3. Germany

In 2018, Germany announced plans to market quantum technology, investing 650 million euros. In July 2020, the German government announced that it would add 2 billion euros in funding to the EU's original 1 billion euros quantum technology plan, which will strengthen research and development in key technologies such as quantum computers, quantum communications, and quantum measurement. It is hoped that Germany will develop its first quantum computer within five years, establish a quantum ecosystem, develop cutting-edge applications, and use superior quantum computing capabilities to enhance Germany's advantages in various industries. In June 2021, IBM and Germany together announced the German Quantum Computing Plan to build Europe's first quantum computer in the IBM Computer Center near Stuttgart, Germany. This will be the most powerful quantum computer in the European industrial field. IBM announced that it will set up the first quantum data center in Europe, which will be operational in 2024. It will support multiple IBM quantum computing systems. Each system is equipped with a "utility-scale quantum processor" and the overall computing scale exceeds 100 qubits. The data center is to be built at the IBM campus in Ehningen, Germany, and will become the IBM Quantum European Cloud Region. Currently, more than 60 corporate members of the IBM Quantum Network access quantum hardware and software through the cloud, including German industrial giant Bosch, Germany's large particle physics laboratory Deutsches Elektronen-Synchrotron, the European Organization for Nuclear Research, and Fraunhofer-Gesellschaft. Ten major companies including BMW in Germany have jointly established the Quantum Technology and Application Consortium (QUTAC), with the goal of developing quantum computing into a real industry. The initial focus is on technology, chemistry, pharmaceuticals, insurance, and automobile industries. In 2021, Germany's "Munich Quantum Valley" was launched, establishing quantum computing as the future, focusing on developing quantum computing technology, secure communication methods, and basic research on quantum technology. Another Quantum Valley was also announced in Lower Saxony.

A4. France

The French government established a national quantum technology strategy in January 2020. In January 2021, French President Emmanuel Macron invested 1.8 billion euros in the five-year quantum technology plan, significantly increasing the government's investment in quantum technology from 60 million euros per year to 200 million euros. By investing such a large amount in the development of quantum computers, quantum sensors, and quantum communications, France hopes to surpass the United Kingdom and Germany and become one of the world's top quantum powers. The French government has invested in several start-up companies, for example, Quandela, which teams up with CEA-Leti for the development of quantum photonic chip, and Paris-based PASQAL, which aims to commercialise a 1,000-qubit quantum computer by 2024.

A5. United Kingdom

In recent years, there has been a growing amount of quantum research in the United Kingdom. In 2015, the UK invested more than 385 million pounds in the first five-year phase of quantum research, mainly researching and developing sensitive gravity detectors, quantum simulators, quantum computers, and quantum clocks. The National Quantum Computing Center was established in 2018 to build quantum computers. At the end of 2019, in the second five-year period, it was announced that the UK's first commercial quantum computer will be manufactured in Abingdon, developed with the assistance of Rigetti Computing and in partnership with Oxford Instruments, Standard Chartered, quantum software company Phasecraft, and the University of Edinburgh. To date, the UK has invested more than £1 billion in two stages of quantum technology development. Oxford quantum computer announced in 2021 that it will cooperate with

Cambridge Quantum Computing to launch Quantum Computing as a Service (QCaaS). The National Quantum Strategy that was published at 2023 is a 10-year vision and action plan for the UK to become a leading quantum-enabled economy, recognising the importance of quantum technologies for the UK's prosperity and security. Government investments to date through the National Quantum Technologies Programs have given the UK a world-leading position.

A6. Netherlands

The Netherlands' strength in the field of quantum technology is demonstrated by its dominance in scientific publications. With only just 17 million people, it is third in the world for scientific citations in quantum research. Quantum research in the Netherlands is concentrated in Delft, Amsterdam, Leiden, Eindhoven, and Twente, all with their own affiliated universities and research centers. The Netherlands has invested approximately 350 million euros in the quantum field in the past five years. For example, the initiative Quantum Delta aims to promote cooperation among multiple research hubs in the Netherlands. The Netherlands began to develop into a quantum knowledge cluster as early as 2010, and quantum technology has long been the only direction of choice for most technology multinational companies. Start-up companies such as QuSoft and QuTech have achieved considerable performance and scale. Microsoft launched the Microsoft Quantum Laboratory in cooperation with the Delft University of Technology (TU Delft) in 2019. Intel, LG, and Fujitsu have also cooperated with the Delft University of Technology on research and development of different quantum projects.

A7. Finland

Finland is in the frontline of the quantum revolution in the EU. Currently, the academy-funded Centre of Excellence in Quantum Technology brings together the scientific and technological knowledge in the field and the cutting-edge research infrastructure. Possible areas of application include quantum sensors, simulators, communication, and computing, which can offer groundbreaking scientific, economical, and societal benefits. Finland's IQM

Quantum Computers is the leader in quantum computers in Europe. It is a new technology start-up company established by a joint technological partnership between Aalto University and VTT in Finland. The current fundraising scale has exceeded 71 million euros, and it is building Finland's first commercial 54-superconducting qubit quantum computer. The Q-Exa alliance led by IQM is also building a quantum computer in Germany. Bluefors is the industry standard for ultra-low-temperature cooling solutions used in quantum technology and its cryogenic technology empowers companies to solve the world's greatest challenges in manufacturing qubits.

A7. Russia

The National Quantum Action Plan was announced in 2019, with an investment of approximately US$790 million within five years to build a practical quantum computer, hoping to catch up with other countries in the field of practical quantum technology. Russia lags far behind in developing and manufacturing classical computer components, but new quantum technologies provide a once-in-a-lifetime opportunity to overtake. Quantum computing and quantum artificial intelligence are two possible breakthrough technology directions.

A8. Japan

The total investment in quantum technology in Japan is approximately 30 billion yen (approximately US$280 million). The Japanese government launched the Quantum Leap (Q-LEAP) programme in 2018, investing in three research and development directions related to quantum technology: quantum simulation and computing, quantum sensing, and ultrashort pulse laser. The government is expected to invest approximately 1.5–2 billion yen in the Moonshot project, with the goal of developing a fault-tolerant universal quantum computer by 2050. In July 2021, Japan and IBM jointly announced the installation and launch of IBM quantum computers at the Kawasaki Business Incubation Center (KBIC) in Japan. In 2021, 25 Japanese companies established the Quantum Strategic industry Alliance for Revolution: Q-STAR. Fujitsu and Riken plan to build two 64-qubit quantum computers and will soon reach 1,000 qubits.

A9. South Korea

South Korea launched the Quantum Computing Technology Development Project in 2019, with a five-year funding of US$40 million. The strategic plan is to foster basic research and development of applications by 2030 so as to encourage industries to enter the quantum business realm. It includes four pillars of core strategy. The first pillar is to focus on high-risk fundamental research, which will be the basis for innovation. Workforce development is the second pillar, which can never be overemphasised. Quantum science and technology is a field that needs multidisciplinary talents. Its root is in the physics and mathematics, but it needs experienced professionals from other fields, including electronics, computer science, and software. The plan's third pillar is to build up the necessary infrastructure such as fabrication facilities and testbeds for components, modules, systems, and applications. The fourth pillar is the applications towards the market. Industry involvement is crucial for the success of the quantum investment. South Korea will pour more than 3 trillion won (US$2.3 billion) into quantum science and technology by 2035, aiming to become the world's fourth-largest powerhouse in the sector. The capital allocation is more than tenfold the quantum technology funding in the country between 2019 and 2023. South Korea makes quantum moves bidding to boost its capabilities to 85% of US levels by 2035. PASQAL forms a quantum partnership in Korea with KAIST and Daejeon City to nurture a leading quantum ecosystem in the Daedeok Quantum Cluster. Yonsei University and Seoul National University are collaborating with IBM and other top universities worldwide to advance quantum education in Korea. South Korea will also dispatch 500 local experts overseas every year.

A10. Australia

Australia has been a leader in quantum technology research and development for almost 30 years, and has produced over 2,500 PhDs in that time. Across the nation, there are world-leading teams working on silicon quantum computers, quantum communications, machine learning, algorithms, and quantum devices. Australia has the capability to translate these scientific advantages into commercial products and services. The National Quantum Strategy of Australia wishes to build a quantum workforce and infrastructure capability while supporting and collaborating with local and international partners. The hardware produced in Australia includes various qubits and control systems, while the software applications include quantum technology in machine learning, medical diagnosis, signal and data processing, financial applications, and self-driving cars. In June 2022, an Australian team developed a quantum integrated circuit composed of several atoms. By precisely controlling the quantum states of atoms, the new processor can simulate the structure and properties of polyacetylene molecules, potentially unlocking new materials and catalysts in the future.

A11. Taiwan

Quantum computers are regarded as the computing tools for the next generation, and the government of Taiwan plans to invest NT$8 billion (US$288.2 million) from 2022 to 2026 to create an industry cooperation platform, which will allow Taiwan to play a leading role among new-generation quantum technology competitors and create the next semiconductor miracle. Quantum technology is expected to have a revolutionary impact on numerous types of applications including in cybersecurity, financial technology, industry, transportation, and national defence. The technology required for quantum computing aligns with Taiwan's existing high-tech industry strengths in chip manufacturing processes, peripheral cables, and packaging technology. Taiwan already has the relevant technologies. The government has launched a cross-agency plan to upgrade Taiwan's hardware, software, and development capabilities for quantum technology.

A12. Singapore

Singapore established the National Center for Quantum Technology (CQT) in 2007 to conduct quantum technology research. In recent years, the research results have been increasingly applied for commercial purposes, and many quantum technology start-ups have emerged. Singapore's neutrality allows countries to exchange quantum technologies, playing a very important role in the development of quantum technology in the world. There are currently about 200 scientific researchers in CQT, which has accelerated the output of excellent scientific research. Singapore is the leading country for quantum technology research and development in Asia.

The National Research Foundation (NRF) of Singapore has set up a special department to convert quantum technology into commercial products.

A13. Israel

Israel has a strong foundation for basic research in quantum mechanics. So far, Israel has focused its main efforts on developing quantum programmes and software, and has accessed the quantum computers in the United States or elsewhere through cloud connections. Israel has only recently begun to develop quantum computers and other hardware. It is estimated that NIS 1.4 billion (approximately US$420 million) will be invested in quantum technology projects in the future. Around 30 Israeli quantum computing companies were funded in the last few years and many have raised a lot of money.

A14. BRICS and Emerging Countries

Since quantum technology is an emerging technological industry for most countries around the world, everyone is almost on the same starting line. Therefore, many countries that have not caught up with the semiconductor technology revolution are actively conducting research in quantum technology, hoping to participate in the "Second Quantum Revolution." Although countries in Africa and Latin America are also trying to form a "Quantum Africa" or "Quantum Latin America" to participate in quantum technology, the countries with most advancements in related fields are Brazil, Russia, India, China, and South Africa, representing the BRICS, and Thailand, representing the emerging countries. The BRICS countries announced at the end of 2020 that they would jointly develop quantum communications with the Russian Rostec National Company. South Africa's Wits University became IBM's first African academic partner and it also collaborated with 15 other universities. India's finance minister announced in 2020 that 80 billion rupees would be allocated over five years to the National Mission on Quantum Technologies and Applications. In 2021, Calcutta University and IIT Kharagpur become IBM's quantum computing academic partners. India is taking significant steps towards establishing itself as a leading player in the global quantum computing industry. With the right support and investment, India has the potential to become a major hub for quantum computing research and development.

Appendix B

The Progress of Quantum Computing for Major Companies

B1. Google

In 2013, Google collaborated with D-wave to develop the 512-qubit D-Wave Two quantum computer. Later, Google and NASA collaborated to develop the 72-qubit superconducting qubit chip Bristlecone. In October 2019, Google announced in Nature magazine that it had achieved "quantum supremacy." The Google team used a 53-qubit quantum computer, Sycamore, to sample a quantum circuit 1 million times in 200 seconds, completing a computing task that would take the supercomputer Summit 10,000 years to complete. Google CEO Sundar Pichai said that this experiment is actually not practical but has a significance similar to that of the Wright brothers' first airplane, which convinced everyone that it was possible for an airplane to fly into the sky. In September 2020, Google was once again featured in "Nature" magazine for its research on Sycamore. It used 12 qubits to simulate the chemical reaction pathway of a diazene molecule with 2 nitrogen atoms and 2 hydrogen atoms. This research pioneered the simulation of quantum chemistry and provided a quick way to develop new molecules and materials. In January 2021, German pharmaceutical company Boehringer Ingelheim announced its cooperation with Google's quantum AI department to use quantum computing for disease research and new drug development. By combining

255

Google's quantum computers and algorithms and Boehringer Ingelheim's pharmaceutical design and modeling experience, they aim to accelerate drug molecular dynamics simulations using quantum computing. In May 2021, Pichai said that he would establish a quantum artificial intelligence park in California, invest billions of dollars in quantum computers, and launch a commercial quantum computer with one million qubits by 2029.

B2. IBM

In May 2016, IBM launched a five-qubit quantum computer and the Quantum Experience cloud free service. IBM officially created a superconducting qubit prototype at the 2018 US International Consumer Electronics Show (CES), which was called a contemporary work of art by visitors due to its beautiful appearance. On January 8, 2019, IBM exhibited the first commercial quantum computer, IBM Q System One. In September 2020, IBM released the IBMQ Hummingbird 65-qubit quantum computer and also released a blueprint for the development of future quantum computers. IBM launched the 127-qubit processor "IBMQ Eagle" in 2021, the 433-qubit processor "IBMQ Osprey" in 2022, and the 1,121-qubit processor "IBMQ Condor" in 2023. IBM further updated its quantum roadmap in December 2023, releasing IBM Quantum System Two and Quantum innovation roadmap to 2033. IBM Quantum System Two begins operation with three IBM Heron processors, designed to bring quantum-centric supercomputing to reality. To prepare for today's era of quantum utility and the coming era of quantum-centric supercomputing, IBM claimed to achieve the quantum computer with error correction in their new quantum computer roadmap before 2030. In order to meet the needs of modular quantum computing, chip-level short-range couplers will be launched to closely connect quantum chips. Moreover, quantum communication between quantum processors must be established to complete a large-scale quantum computer. By building modular and scalable processor clusters, IBM will launch systems with more than 4,000 qubits in 2025, and even exceed one million qubits after 2025. Logical memory qubits will be demonstrated in 2026, followed by logical communication among qubits the following year. Logical gates will be available in the Starling processor in 2028, and IBM expects to have the full package working on an iteration of Starling in 2029. This advancement will allow a significant increase in quantum gate operations, from 15,000 gates on Flamingo in 2028 to 100 million gates on Starling in 2029.

IBM's ultimate goal is to combine quantum computers and supercomputers into a system to solve problems in defence, finance, medicine, and pharmaceuticals. In April 2021, IBM announced a quantum computer cooperation plan with the Cleveland Clinic, aiming to accelerate the development of epidemic-related biomedical technologies using the high-speed computing capabilities of quantum computers, thereby preventing future pandemics. In addition to selling a quantum computer to the Cleveland Medical Center, IBM also sold at least another five quantum computers, one to Germany, another to the University of Tokyo in Japan, one to South Korea, and one each to Canada and Spain. As of 2022, more than 180 units, including academic institutions, national laboratories, start-ups, universities, and enterprises, have used IBM's quantum computer platform through IBM Quantum Experience. IBM has also announced that it will pursue the quantum performance of quantum computers in the future: including more qubits, higher quantum volume, and better computing speed. Calculation speed is the number of quantum gate operations per second, which is CLOPS (Circuit Layer Operations Per Second). In 2024, IBM offered a tool capable of calculating unbiased observables for circuits with 100 qubits and depth-100 gate operations in a reasonable run-time. Thanks to Heron with error rates below the "three-nines" gate fidelity threshold, plus the software infrastructure to read out the circuits in concert with classical resources, a circuit of 100×100 can be run in less than a day and produce unbiased results. This system will be able to run quantum circuits with complexity and runtimes beyond the capabilities of the best classical computers today.

B3. Rigetti Computing

Rigetti Computing specialises in developing hardware and software for quantum computers. Rigetti's first product, Forest, was released in 2017. It has been used by Oak Ridge National Laboratory to simulate the structure of deuterium nuclei and has conducted more than 65 million quantum computing experiments. Rigetti Computing received

US$79 million in Series C financing in 2020, with a total funding of approximately US$198.45 million. Rigetti Computing will also work with the UK to develop the UK's first quantum computer. Rigetti was officially listed on the Nasdaq stock exchange in the United States through SPAC in March 2022. In 2023, Rigetti's 84-qubit Ankaa-2 quantum system was made publicly available to all of its customers.

B4. Quantinuum (Honeywell)

In October 2020, Honeywell used trapped ion technology to complete a quantum computer with 10 fully connected qubits — Model H1. Honeywell also announced its quantum computing blueprint for the next 10 years, with the intention of launching a 64-qubit ion trap quantum computer within 3 years. At the end of 2021, Honeywell Quantum Solutions merged with Cambridge Quantum. The new company was named "Quantinuum" and became the world's largest quantum computing company. Quantinuum H-Series hardware demonstrated excellent performance improvement in 2023. Quantinuum's quantum computer Model H1-1 has demonstrated a quantum volume of 524,288, which is 1000× higher than the next best reported quantum volume. Quantinuum was already valued at US$5 billion in 2024 and secured new US$300 million in equity funding.

B5. IonQ

IonQ was founded in 2015 by the University of Maryland and its collaborators, with US$2 million in seed funding. The company successfully built two ion trap quantum computers in 2018 and launched a 32-qubit quantum computer in 2020. IonQ, a quantum start-up company, was established in 2015 and officially announced its listing on the New York Stock Exchange through SPAC in March 2021. In just 6 years, it went from a start-up company with a market value of 2 million to a unicorn with a market value of US$2 billion. It is the world's first publicly listed start-up company in quantum computing. However, in May 2022, Scorpion Capital publicly claimed that there was a problem with IonQ's 32-qubit quantum computer, causing IonQ's stock price to plummet rapidly. The current doubts about about the 32-qubit system still need to be clarified, but it shows that there is still a huge cognitive gap between the technological achievements of quantum start-ups and the public's perception of commercialisation.

IonQ introduced Reconfigurable Multicore Quantum Architecture (RMQA) technology in 2021. There are 4 16-ion trap chains on a single chip. Each chain can be dynamically configured into the quantum computing core, laying the foundation for achieving 1,000 ions on a single chip and enabling parallel multicore quantum processing. IonQ achieves excellent quantum computing performance by utilising glass chips, which offer transparency and help avoid the electric fields associated with silicon-based chips. IonQ Forte, built in 2023, is the world's first software-configurable quantum computer. It uses ytterbium ions and integrates highly specialised acousto-optic deflectors (AODs) to direct laser beams at individual qubits in the ion chain to apply logic gates among the qubits. This approach provides unprecedented precision and stability to the laser beams, enhancing fidelity and reliability by minimising noise and unintended residual light on neighboring qubits.

B6. Intel

In February 2020, Intel released Horse Ridge, a control chip designed for ultra-low-temperature environments for quantum computing. It simplifies the current complex control electronic equipment while optimising qubits and increasing the scalability of quantum systems. The technology currently proposed by Intel can effectively isolate electronic components at around 1.1 Kelvin from qubits at 10^{-2} Kelvin. Therefore, the process of integrating qubits and control electronic components on the same chip becomes simple. This is an indispensable key technology for million-qubit quantum computers. In April 2022, Intel announced that it would cooperate with QuTech in the Netherlands to form a quantum research unit. Leveraging Intel's factory in Oregon, USA, they successfully manufactured qubits on 300mm silicon wafers. The new process uses advanced semiconductor manufacturing technologies, including all-optical lithography technology used to produce silicon spin qubits, and those in the production of Intel's latest generation complementary metal oxide semiconductor (CMOS) chips, which are all current factory equipment and wafers. The yield rate has exceeded 95%. In 2023, Intel announced its newest quantum research chip, Tunnel Falls, a 12-qubit silicon chip, which is available to the quantum research community.

B7. AMD

AMD introduced an improved version of the traditional multi-SIMD (Single Instruction Multiple Data) method in 2021, using quantum teleportation to improve the reliability of the quantum system and reducing the number of qubits required for calculations. Quantum states are very sensitive to the environment and even the slightest influence may cause decoherence. The sensitivity of a quantum system tends to increase as more qubits are added to the system. AMD's design focuses on linking qubits across regions, allowing work that theoretically needs to be performed sequentially to be executed in an out-of-order manner. This approach could improve the scalability of quantum computers and bring them closer to the goal of commercial quantum computers.

B8. Microsoft

Microsoft has been developing quantum computers based on topological qubits. Topological qubits are immune to variations in the environment, reducing error correction procedures. But there are currently many technical bottlenecks. In order to reduce R&D risks, Microsoft has begun to cooperate with IonQ, Honeywell, and QCI. It also cooperates with the Microsoft Azure Quantum platform. Developers can use Microsoft's quantum development tool set to write and execute quantum computing programmes. Microsoft has been sponsoring Leo Kouwenhoven, a professor at the Delft University of Technology in the Netherlands, and Charles Marcus, a professor at the University of Copenhagen, to study topological qubits. However, in 2021, Professor Kouwenhoven claimed that the experimental results in the "Nature" article in 2018 were wrong and could not prove the existence of topological quantum states. In 2022, Microsoft's Azure Quantum team once again claimed that it could produce topological superconducting phases and their accompanying Majorana zero modes, breaking through important obstacles in building scalable quantum computers. In 2023, Microsoft announced its roadmap with topological qubits, on which the company's researchers have been working for several years. Microsoft believes that it will take fewer than 10 years to build a quantum supercomputer using these qubits, and the computer will be able to perform a reliable one million quantum operations per second.

B9. PsiQuantum

Quantum start-up PsiQuantum focuses on photonic quantum computing. In May 2021, PsiQuantum and wafer foundry GlobalFoundries announced that they would work together to build PsiQuantum's Q1 system, which is the world's first quantum computer with more than 1 million qubits. This system is supposed to demonstrate that semiconductor manufacturing processes can produce the silicon photonic and electronic chips required for quantum computers on silicon wafers. This is a major breakthrough in the cooperation between quantum and semiconductor technologies. In July 2021, PsiQuantum announced that it had raised US$450 million in Series D funding to produce the next generation of quantum computers. Since 2016, it has raised a total of US$665 million.

B10. Xanadu

Xanadu is a Canadian quantum company established in 2016 with the purpose of developing photonic quantum computers. In 2020, it announced the successful development of a programmable photon quantum chip and officially provided 8-qubit and 12-qubit photonic quantum computers for public use. Xanadu and Korea Institute of Science and Technology Information (KISTI) partnered in 2023 to create South Korea's first quantum-classical hybrid computing infrastructure.

B11. QuEra

QuEra is a start-up company formed by the Harvard-MIT Center for Ultracold Atoms. It has developed a cold-atom quantum computer called a "programmable quantum simulator." This special type of quantum computer can operate with 256 qubits, and the number of quantum states that can be expressed exceeds the number of atoms in the solar system. QuEra will introduce a third-generation quantum error-corrected model with 100 logical qubits and over 10,000 physical qubits in 2026. This development, capable of deep logical circuits, will push quantum computing beyond the limits of classical simulation, ushering in a new era of discovery and innovation. QuEra has announced that a quantum computing testbed, funded by the UK's National Quantum Computing Centre (NQCC), will be built by QuEra with UK-based collaborators. QuEra will begin work on its testbed for the NQCC soon and expects it to be operational in early 2025.

B12. ColdQuanta

ColdQuanta leverages cold-atom technology for quantum computing, sensing, and networking. The company's technology focuses on improving positioning and navigation systems. ColdQuanta provides products for cold-atom experiments, quantum simulations, quantum information processing, atomic clocks, and inertial sensing. ColdQuanta provides the cold-atom quantum computer, Hilbert, in the cloud. The team aims to reach 1,000 qubits soon, with strong connectivity, high fidelity, and miniaturisation at room temperature.

B13. PASQAL

The Paris-based start-up PASQAL was founded on the research of Nobel Prize-winning co-founder Alain Aspect. PASQAL has already sold two 100-qubit quantum computers for installation in Germany and France. But the company has higher goals. PASQAL announces a new roadmap focusing on business utility and scaling beyond 1,000 qubits towards fault tolerance era starting in 2025. It aims to reach 10,000 qubits by 2026 through a scalable architecture for logical qubits. To fund that work, it raised €100 m in a Series B round led by Temasek. PASQAL also offers cloud access to its quantum computers located in Paris and is exploring potential quantum computing applications, working in partnership with several Fortune 500 companies, such as BASF, Siemens, Airbus, and Crédit Agricole CIB.

B14. IQM Quantum Computers

Another top player in Europe's vibrant quantum ecosystem, IQM is a collaborative initiative by Aalto University and VTT Technical Research Centre of Finland. Its most sophisticated computer has 54 qubits but it also sells less powerful, five-qubit quantum processing units to universities and research labs. Its processors will also be used in the first quantum accelerator in Spain. IQM, which raised €220.5 m in its first five years of existence, employs 280 people across its Espoo headquarters and offices in Paris, Madrid, and Munich. IQM introduced a 5-qubit IQM Spark processor for universities and research labs in 2023.

B15. CIQTek (国仪量子)

CIQTek, headquartered in Hefei and led by academician Du Jiangfeng of the Chinese Academy of Sciences, makes quantum precision measuring instruments. Instruments such as the quantum diamond single-spin spectrometer, the quantum diamond atomic force microscope, and the diamond educational quantum computer are all globally pioneering exclusive products from CIQTek due to which they face less competition in the market. On the other hand, the main market for products such as electron paramagnetic resonance spectrometers and scanning electron microscopes is sales within China.

B16. QuantumCTek (国盾量子)

QuantumCTek Co., Ltd., established in 2009 with its original techniques developed at the University of Science and Technology of China, is a pioneer, practitioner, and leader of the commercialisation for quantum information technology. QuantumCTek is committed to providing competitive products and services for quantum secure communication, quantum computing, and quantum measuring. In 2020, QuantumCTek was listed on the SSE STAR Market, becoming the first A-share listed company with quantum technology as its core business. QuantumCTek, one of the few companies world-wide with full capacity to design, manufacture, and deploy quantum-secure communication networks, has been actively promoting the construction of an "integrated space & ground" quantum communication network for the Yangtze River Delta and the whole of China. Products from QuantumCTek have been deployed in many backbone networks including the quantum-secure communication "Beijing-Shanghai Trunk Line," metropolitan area networks, and access networks, serving customers in government, finance, and other industries. QuantumCTek has been conducting extensive research in the Internet of Things, big data, artificial intelligence, and 5G, collaborating with ecological partners and providing "quantum+" security services to various industries. In the quantum computing field, QuantumCTek has been participating the quantum computing superiority experiment on the superconducting quantum computer "Zu Chongzhi", (祖冲之) and is capable of providing complete solutions for superconducting quantum computing. Its cloud platform service, based on a quantum computer similar to the 176-qubit superconducting quantum computer "Zu Chongzhi," has been launched online and is open to global users.

B17. Origin Quantum (本源量子)

Origin Quantum is China's first innovative company to develop, promote, and apply quantum computers, and has successfully delivered the first quantum computer to users. Origin Quantum has developed multiple quantum computers, quantum chips, integrated quantum measurement and control machines, and has launched the Origin quantum computing cloud platform, a self-developed quantum computer operating system, the quantum programming software development tool QPanda, quantum finance applications, and other products. Recently, Origin Quantum has developed a dilution refrigerator Origin-SL400, a core component of quantum computers. It can provide an ultra-low-temperature environment below 12 mK and a cooling power of no less than 400 μW at 100 mK, and is available for global sale. Origin Wukong, named after the Monkey King from Chinese mythology, represents China's leading entry into the third generation of superconducting quantum computers with 72 qubits. It was made accessible to global users on January 6, 2024, and since then, it has attracted remote visitors from 61 countries, with American users leading the tally.

B18. SpinQ (量旋量子)

Founded in 2018, SpinQ Technology Inc. is a one-stop solution provider dedicated to the industrialisation and popularisation of quantum computing. With a strategy driven by both technology research and commercial implementation, SpinQ has established a comprehensive industry layout through superconducting quantum computers, NMR (Nuclear Magnetic Resonance) quantum computers, a quantum computing cloud platform, and application software. SpinQ is empowering various cutting-edge fields such as scientific research, education, drug development, fintech, and artificial intelligence, collaborating with partners to build scenario-based solutions that integrate quantum computing into diverse industries, making it a true productivity tool.

In the field of NMR quantum computing, SpinQ has achieved groundbreaking innovations in miniaturising quantum computers, leading the industry in commercialisation. Its achievements include the world's first desktop quantum computer, "Gemini," and the first portable quantum computer, "Gemini Mini." SpinQ has achieved widespread sales and established an absolute leading position globally.

In the field of superconducting quantum computing, SpinQ continuously improves the technology research and development, ranging from superconducting quantum chips, quantum chip electronic design automation (QEDA), superconducting quantum measurements, and control systems to complete superconducting quantum computing systems. Through an integrated approach across the entire value chain, SpinQ is advancing towards industrial applications.

In addition, SpinQ has also launched the quantum software programming framework SpinQit and the quantum computing cloud platform "Taurus," providing one-stop solutions from software and hardware products to application services.

B19. QBoson (玻色量子)

QBoson is a photonic quantum start-up company established by professionals graduating from Stanford University. In 2023, Qboson commercially produced the CIM (Coherent Ising Machine), a coherent quantum

AI coprocessor. The Coherent Ising Machine uses coherent light pulse phase encoding and mutual injection between optical pulses to realise the training of quantum neural networks and solve complex combinatorial optimisation problems.

B20. TuringQ (图灵量子)

TuringQ was founded by Professor Jin Xianmin, who graduated from the University of Cambridge and currently works at Shanghai Jiao Tong University. The company demonstrated that lithium niobate thin film (LNOI) photonic chips can realize photonic circuits with three-dimensional integration and programmable functions. TuringQ aims to realise a universal optical quantum computer with the ability to manipulate millions of qubits, while maintaining low environmental requirements and high integration capability.

B21. Qudoor (启科量子)

Qudoor focuses on the manufacturing of quantum communication equipment and the development of quantum computers. It launched the research of distributed ion trap quantum computers in 2020 and developed China's first ion trap quantum computer AbaQ-1 in 2023. Its goal is to create a hundred-qubit distributed ion trap quantum computer with the quantum volume reaching 100 million. In terms of quantum computing applications, Qudoor has established cooperative relationships with leading companies in insurance, securities, new drug development, encryption and decryption, and other fields to build a market ecosystem.

B22. Hyqubit Quantum (华翊量子)

Hyqubit Quantum focuses on ion trap quantum computing. The founding team is made up of graduates from the Quantum Information Center of Tsinghua University and is led by Professor Duan Luming. The company plans to create commercial quantum computers with more than a hundred qubits in the near future and gradually increase the number of qubits to thousands or even tens of thousands in the next few years.

B23. Fujitsu

At present, quantum computers have problems such as qubit errors, and it is difficult to perform complex calculations accurately, but quantum simulators can perform these tasks without errors. Fujitsu, based in Japan, uses a hybrid quantum computing platform to achieve optimal quantum computing and manage acceleration by connecting a 64-qubit superconducting quantum computer and a 40-qubit quantum simulator. Utilising the "Computing Workload Broker" technology that is under development, optimal quantum computing is achieved by automatically combining various computing resources and algorithms. Fujitsu has also developed the Digital Annealer using semiconductor components to optimise the processing of Ising-like problems.

B24. NEC

NEC from Japan is aiming to establish practical applications for quantum annealing machines, which are a type of quantum computer. Quantum annealing machines can solve combinatorial optimisation problems that were previously difficult to solve due to the need for massive amounts of computation. In addition, while awaiting the realisation of high-performance quantum annealing machines, NEC is developing annealing simulators based on vector computers that can handle large-scale combinatorial optimisation problems.

B25. NTT

Compared with traditional quantum computing, optical quantum computing benefits from photons being very stable and less susceptible to interference. Optical quantum computing may also be more scalable than superconducting quantum computing because photons can be easily manipulated and transmitted over long distances using optical fibres, compared with signals sent over superconducting wires. This could allow for the future construction of large-scale quantum networks. The Japanese government invested 200 billion yen(~US$1.34 billion) in Japan to develop optical quantum computers and start production of actual machines in 2022. They hope to realise a fault-tolerant universal quantum computer by 2050, which will completely change the economy, industry, and security. NTT has been exploring cutting-edge computing systems leveraging hybrid principles of quantum and classical computing, and it believes that the Coherent Ising Machine (CIM) is the most promising next-generation solution to date.

B26. Foxconn Group

Foxconn Group's Hon Hai Research Institute has established the "Trapped Ion quantum computing Laboratory," aiming to launch a 5- to 10-qubit trapped ion quantum computer within five years as a prototype for mid- to long-term scalable quantum computers.